A Course on the Web Graph

A Course on the Web Graph

Anthony Bonato

Graduate Studies
in Mathematics

Volume 89

American Mathematical Society

Atlantic Association for Research
in the Mathematical Sciences

AARMS

2000 *Mathematics Subject Classification.* Primary 05C90, 05C80, 68R10, 94C15.

For additional information and updates on this book, visit
www.ams.org/bookpages/gsm-89

Library of Congress Cataloging-in-Publication Data

Bonato, Anthony, 1971-
A course on the Web graph / Anthony Bonato.
p. cm. — (Graduate studies in mathematics ; v. 89)
Includes bibliographical references and index.
ISBN 978-0-8218-4467-0
1. Internet—Mathematical models. 2. Telecommunication—Traffic—Mathematical models. 3. Representations of graphs. 4. Information visualization. I. Title.

TK5105.888.B667 2008
004.67′8–dc22   2007060579

∞ The paper used in this book is acid-free and falls within the guidelines
established to ensure permanence and durability.
Visit the AMS home page at http://www.ams.org/

10 9 8 7 6 5 4 3 2 1   13 12 11 10 09 08

Contents

List of Figures

Preface

The internet affects many aspects of our lives, such as how we store and retrieve information, conduct business, and communicate. For example, information is no longer only stored in printed form, but is represented on-line via a complex set of interconnected web pages. The *web graph* has vertices representing web pages, with edges corresponding to the links between pages. The web graph is a real-world network which has undergone intensive study in the last decade by theoreticians and experimentalists. Does this graph have interesting properties? Are there good, rigorous mathematical models for these properties? Can we exploit the graph structure of the web to help search it for information? The answer to all three questions is, of course, yes!

The study of the web graph, or *internet mathematics* as it is now often called, is an active field of study. As the subject is new, there is often a lack of consensus on the central topics, models, even notation, with key questions not always evident. As the subject is fast-breaking, a large arsenal of techniques are required to model and analyze properties of the web. However, possessing the right mathematical tools and a familiarity with current research developments is an important first step. This book should supply a solid mathematical introduction to internet mathematics, and will encourage interest in an emerging and fascinating area of graph theory and theoretical computer science.

The book resulted from lecture notes for an Atlantic Association for Research in the Mathematical Sciences (AARMS) Summer School graduate course *Massive Networks and Internet Mathematics* taught in July 2006 at Dalhousie University in Halifax. A version of the course was taught twice before at Wilfrid Laurier University in Waterloo. As such, the book is appropriate for graduate students or keen undergraduate students in mathematics, computer science, engineering, or physics, whose background includes elementary graph theory, linear algebra, and probability theory. The text is also useful to professional mathematicians, scientists, or engineers interested in learning more about the web graph and graph theory in general. We emphasize that our view is clearly on the mathematics surrounding the web graph. Further, the topics covered are by no means exhaustive.

The book is largely self-contained, and references are given where proofs are omitted. There are over 100 exercises at the end of the chapters and many worked examples, all making the book suitable for either a course or for self-study. Open problems are stated in the exercises and elsewhere.

The book consists of seven chapters. Chapter 1 supplies the requisite background and notation in graph theory and discrete probability used throughout the remaining chapters. We describe the graph and probability theory as well as notation that acts as the foundation for the remaining chapters. The web graph and its key properties are introduced in Chapter 2. Here the reader will learn, among other things, about power law degree distributions and the small world property. Various real-world, self-organizing networks, ranging from technological, biological, to social, are discussed in this chapter. In Chapter 3, an introduction is given to techniques and properties of the classical $G(n,p)$ random graph. Random graph theory supplies the backbone for much of internet mathematics; the techniques used here will be used in later chapters. Chapter 4 surveys the mathematics of stochastic web graph models. Several models are reviewed and analyzed for their degree sequence and other parameters. The topic of searching the web is presented in Chapter 5, where the key web ranking algorithm PageRank—used by the search engine Google—is described. The chapter includes a discussion of the linear algebra and Markov chains used in modern web ranking algorithms. In Chapter 6, we describe the interaction between infinite graph theory and web graph models. The view of massive real-world networks as infinite graphs is relatively new, and it ties in well with the existing theory on the infinite random graph. There are myriad facets to research on the web graph, so as a result we finish in Chapter 7 with three distinct topics on web graph research: spectra of power law graphs, modelling viruses on the web, and domination in web graph models.

How to read this book? The key chapters for a course in internet mathematics are Chapters 1 to 5, inclusive. Chapters 6 and 7 may be completely or partially omitted in a one-semester course. The topics in those chapters are well suited for reading projects. All chapters contain exercises (some with references), and so the book is well suited for assignments and self-study. A web page will be maintained for the book at

`http://info.wlu.ca/~wwwmath/faculty/bonato/webgraph.html`

which will contain useful links and additional information such as corrections or addenda.

As with any undertaking of this nature, there are many people to acknowledge and thank. A sincere thank you goes out to Jon Borwein, David Langstroth, Ron Fitzgerald, Ina Mette, Natalya Pluzhnikov, and everyone at the AMS and AARMS for their generous support of this project. I would like to thank all the mathematicians I have discussed internet mathematics and graph theory with over the years; in particular, I would like to thank Cathy Baker, Kathie Cameron, Peter Cameron, Colin Cooper, Dejan Delić, Fan Chung Graham, Geňa Hahn, Jon Kleinberg, Jarik Nešetřil, Richard Nowakowski, and Joel Spencer. I especially wish to acknowledge Jeannette Janssen, whose infectious enthusiasm and brilliance attracted me to and kept me fascinated with the subject. Thanks to Dejan Delić, Douglas Hamlyn, Jeannette Janssen, Pawel Prałat, Laleh Samarbakhsh, Changping Wang, and the anonymous referees for their careful reading of early drafts of the book. Without the constant love and support of my family—Doug, Anna Maria, Paul, Lisa—this book would not have been written. I dedicate the book to the memory of my sister Paula.

Chapter 1

Graphs and Probability

Mathematical discoveries, small or great, are never born of spontaneous generation. They always presuppose a soil seeded with preliminary knowledge and well prepared by labour, both conscious and subconscious.

–Henri Poincaré

He who has not first laid his foundations may be able with great ability to lay them afterwards, but they will be laid with trouble to the architect and danger to the building.

–Niccolo Machiavelli

1.1. Introduction

Graphs are the central objects we study, and a number of the results we present on the web graph are probabilistic. This chapter supplies the reader with the necessary background and notation in both graph theory and discrete probability theory to read the remainder of this book. Since both subjects are so expansive, we will focus on material that we will use later in the text. The reader who would like additional background in graph theory is directed to [**31**], [**84**], or [**195**], while a good discrete probability reference is [**122**]. The reader with background in both subjects may safely move on to the next chapter.

We will use the following notation throughout. The set of natural numbers (which contains 0) is written $\mathbb{N}$, while the rationals and reals are denoted by $\mathbb{Q}$ and $\mathbb{R}$, respectively. The cardinality of $\mathbb{N}$ is $\aleph_0$, while the cardinality of $\mathbb{R}$ is $2^{\aleph_0}$. If n is a positive natural number, then we define $[n] = \{1, \dots, n\}$.

The Cartesian product of two sets A and B is written $A \times B$, while their difference is written $A \backslash B$.

As the results we present are often asymptotic, let us formalize some notation. Let f and g be functions whose domain is some fixed subset of $\mathbb{R}$. We write $f \in O(g)$ if

$$\limsup_{x \to \infty} \frac{f(x)}{g(x)}$$

exists and is finite. This is equivalent to saying that there is a constant $c > 0$ (not depending on x) and an integer $N > 0$ such that for $x > N$, $f(x) \leq cg(x)$. We will abuse notation and write $f = O(g)$.

We write $f = \Omega(g)$ if $g = O(f)$, and $f = \Theta(g)$ if $f = O(g)$ and $f = \Omega(g)$. If

$$\lim_{x \to \infty} \frac{f(x)}{g(x)} = 0,$$

then $f = o(g)$ (or $g = \omega(f)$). So if $f = o(1)$, then f tends to 0. We write $f \sim g$ if

$$\lim_{x \to \infty} \frac{f(x)}{g(x)} = 1.$$

Let x and y be non-negative, real-valued variables. We write $\log x$ for the natural logarithm of x (sometimes called $\ln x$). If x is a real number, then $1 + x \leq e^x$. We will sometimes write e^x as $\exp(x)$, especially if x is a complicated expression. If $0 \leq m \leq n$ are integers with $n > 0$, then

$$\left(\frac{n}{m}\right)^m \leq \binom{n}{m} \leq \frac{n^m}{m!} \leq n^m.$$

1.2. Graph Theory

Graphs are beautiful combinatorial objects, whose recorded history stretches back to work by Leonhard Euler, who used graphs to solve the Königsberg bridge problem [**101**]. A *graph* G consists of a non-empty *vertex set* $V(G)$, and an *edge set* $E(G)$ of unordered 2-element sets from $V(G)$. A graph is sometimes called *network*, especially with regards to real-world examples. More formally, we may consider $E(G)$ as a binary relation on $V(G)$ which is irreflexive and symmetric. We often write $G = (V(G), E(G))$, or if G is clear from the context, $G = (V, E)$. The set E may be empty. Elements of $V(G)$ are *vertices*, and elements of $E(G)$ are *edges*. Vertices are occasionally referred to as *nodes* or *points*, while edges are referred to as *lines* or *links*. We write uv for an edge $\{u, v\}$, and say that u and v are *joined* or *adjacent*; we say that u and v are *incident* to the edge uv, and that u and v are the *endpoints* of uv.

The most important graph for us is the *web graph*, where vertices represent web pages, and the edges correspond to links between the pages. We write W for this graph, which is a real-world graph. We will discuss properties of W in much more detail in the next chapter. The graph W is often considered as a directed graph, where edges assume an orientation. We will discuss directed graphs at the end of this section.

Graphs may of course be infinite, but the following caveat holds:

All graphs will be assumed to be finite unless otherwise stated.

We often visualize a graph with a small number of vertices by a drawing such as the famous graph in Figure 1.1, where vertices are points, and edges are curves joining points. While drawing a graph with a billion vertices and edges is unrealistic, the visual component of graph theory is quite attractive, and distinguishes it from many other parts of combinatorics and mathematics.

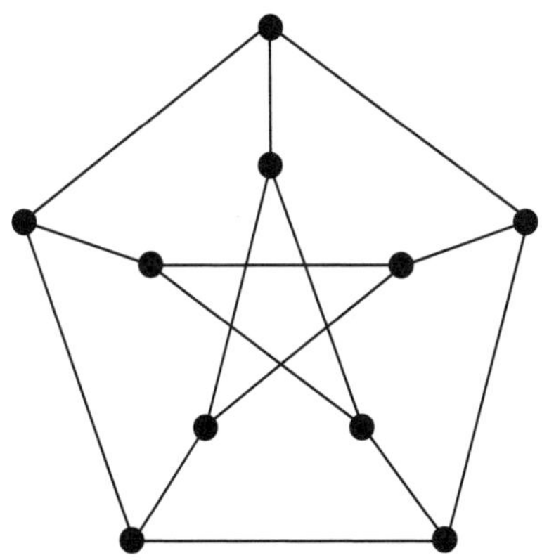

Figure 1.1. The Petersen graph.

The cardinality $|V(G)|$ is the *order* of G, while $|E(G)|$ is its *size*. If G is finite, then

$$|E(G)| \leq \binom{|V(G)|}{2},$$

while if G is infinite, then $|E(G)| \leq |V(G)|$.

On occasion, we may allow graphs to contains loops, or multiple edges. Graphs with neither loops nor multiple edges are *simple*. As a second caveat:

All graphs we consider are simple unless otherwise stated.

Given a vertex u, define its *neighbour set* $N(u)$ to be the set of vertices joined to u (also called *neighbours* of u), while its *non-neighbour set* $N^c(u)$ is the set of vertices distinct from u and not joined to u. Note that $\{u\} \cup N(x) \cup N^c(x)$ partitions $V(G)$.

If $S \subseteq V$, then define the *subgraph induced by* S, written as $G \restriction S$ (or as either $\langle S \rangle_G$ or $G[S]$), to be the graph with vertices S and with two vertices joined in $G \restriction S$ if and only if they are joined in G. A *subgraph* of G is a graph H such that $V(H) \subseteq V(G)$ and $E(H) \subseteq E(G)$; a *spanning subgraph* is a subgraph such that $V(H) = V(G)$. If S is a set of vertices, then $G - S$ is the subgraph induced by $V(G) \backslash S$; if $S = \{x\}$, then we write this as $G - x$. If U is a set of edges, then the subgraph $G - U$ is defined analogously.

A *homomorphism* f between graphs G and H is a function $f : V(G) \rightarrow V(H)$ which *preserves edges*; that is, if $xy \in E(G)$, then $f(x)f(y) \in E(H)$. We abuse notation and simply write $f : G \rightarrow H$. An *embedding* from G to H is an injective homomorphism $f : G \rightarrow H$ with the property that $xy \in E(G)$ if and only if $f(x)f(y) \in E(H)$. We will write $G \leq H$ if there is some embedding of G into H, and say that G *embeds in* H. Note that the relation $\leq$ is an order (that is, it is a reflexive, anti-symmetric, and transitive relation) on the class of all finite graphs, while it is a quasi-order on the class of all graphs (that is, a reflexive and transitive relation). An *isomorphism* is a bijective embedding; if there is an isomorphism between two graphs, then we say they are *isomorphic*. We write $G \cong H$ if G and H are isomorphic. See Figure 1.2 for drawings of isomorphic graphs. For example, for all $n \geq 1$, C_{2n+3} admits a homomorphism to C_{2n+1} but the converse is false.

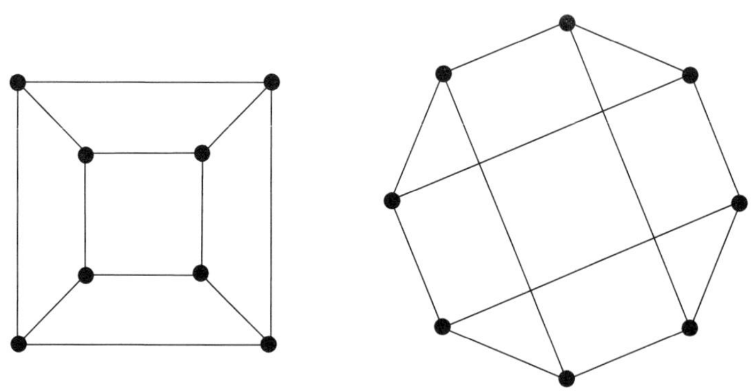

Figure 1.2. Isomorphic graphs.

Isomorphic graphs satisfy all the same properties. Note that if $f : G \rightarrow H$ is an embedding, then the subgraph induced by the image $f(V(G))$ is an induced subgraph of H isomorphic to G. The relation $\cong$ is an equivalence relation on the class of all graphs, whose equivalence classes are *isomorphism types* or *isotypes*. We will always identify a graph with its isomorphism type.

An *automorphism* of a graph G is an isomorphism from G to itself; the set of all automorphisms forms a group under the operation of composition, written $\mathrm{Aut}(G)$. A graph G is *vertex-transitive* if for every pair of vertices u and v there is an automorphism of G mapping u to v. For example, the graphs in the two previous figures are vertex-transitive.

The *degree* of a vertex is the cardinal $|N(u)|$, and is written $\deg_G(u)$ or simply $\deg(u)$. The degree of a vertex is one of the simplest but most useful parameters of a graph. The following theorem, affectionately called the *First Theorem of Graph Theory*, is easy to prove but is important in the development of graph theory.

THEOREM 1.1. *If G is a graph, then*

$$2|E(G)| = \sum_{u \in V(G)} \deg(u).$$

Proof. When summing the degree of each vertex, each edge is counted exactly twice. □

As one corollary of Theorem 1.1, the number of odd degree vertices in a graph is even. The *minimum degree* of G is written $\delta(G)$, while the *maximum degree* (which is the supremum of all degrees if G is infinite) is written $\Delta(G)$. For a cardinal k, a graph is *k-regular* (or just *regular*) if each vertex has degree k. By the First Theorem of Graph Theory, k-regular graphs with k odd exist only for even orders.

A *walk* in a graph consists of an alternating sequence of vertices and edges

$$x_0, e_1, x_1, \ldots, e_t, x_t \tag{1.1}$$

so that for all $1 \leq i \leq t$, $e_i = x_{i-1}x_i$. Note that vertices and edges may be repeated in a walk. To be explicit about the endpoints, we sometimes refer to (1.1) as an *x_0,x_t-walk*. A walk is *closed* if $x_0 = x_t$; otherwise, it is *open*. The number of edges is the *length* of the walk. A *path* is an open walk with no repeated vertex. A *cycle* is a closed walk with no repeated vertex. The path of length n is P_n, and the cycle of length n is C_n. The *girth* of a graph G, written $g(G)$, is the minimum length of a cycle in a path (the girth is ∞ if G is *acyclic*; that is, contains no cycle). As a straightforward illustration of these notions, we prove the following lemma.

LEMMA 1.2. *If G is a graph, then G contains a path of length $\delta(G)$, and if $\delta(G) \geq 2$, then $g(G) \leq \delta(G) + 1$.*

Proof. Let P be a path of maximum length r in G, with consecutively joined vertices $u_0, \ldots, u_r$. As all the neighbours of u_r are in P, we have that $r \geq \delta(G)$.

Assume that $\delta(G) \geq 2$. Let u_m be the vertex joined to u_0 with maximum index, where $m \geq 2$. Then the cycle formed by joining u_0u_m to the path from u_0 to u_m has length at least $\delta(G) + 1$. □

A graph is *connected* if for each pair of vertices there is a path between them. The relation of being connected by a path is an equivalence relation on V, and the equivalence classes are the *connected components* of G. A graph which is not connected is called *disconnected*; a connected component consisting of a single vertex is called an *isolated vertex*. A vertex joined to all other vertices is called *universal*.

The *distance* between u and v, written $d(u, v)$, is either the length of a shortest path connecting u and v (and 0 if $u = v$) or ∞ otherwise. Note that $d(u, v)$ turns each graph into a metric space. The *diameter* of a connected graph G, written $\text{diam}(G)$, is the supremum of all distances between distinct pairs of vertices. If the graph is disconnected, then $\text{diam}(G)$ is ∞. For example, $\text{diam}(C_n) = \lfloor \frac{n}{2} \rfloor$.

A *chord* of a cycle is an edge joining two non-consecutive vertices. An acyclic connected graph is a *tree*, while an acyclic graph is a forest. A *spanning tree* is a spanning subgraph that is a tree.

THEOREM 1.3. *A graph G is connected if and only if it contains a spanning tree.*

Proof. As the reverse direction is immediate, we prove the forward direction. We inductively construct a tree which contains each vertex of G. Fix u a vertex of G, and let T_0 be the subgraph induced by $\{u\}$. Suppose that T_n is a tree of order n containing T_0, where $n \leq |V(G)|$. Suppose that $n < |V(G)|$. Since G is connected, among the vertices not in T_n there is one joined to some vertex z joined to a vertex y of T_n. Add the edge yz to T_n to form the subgraph T_{n+1} of order $n + 1$. The induction continues until each vertex of G is in the tree. □

The analogue of Theorem 1.3 holds for infinite graphs assuming the Axiom of Choice. Trees have several different characterizations such as the ones in the following theorem (whose proof is an exercise).

THEOREM 1.4. *A graph G is a tree if and only if any one of the following properties is satisfied.*

(1) *Every pair of distinct vertices of G is connected by a unique path.*
(2) *The graph G is connected, but deleting any edge disconnects G.*
(3) *The graph G is acyclic, but adding any edge to G forms a cycle.*
(4) *The graph G is connected, and has $|V(G)| - 1$ many edges.*

A *Hamilton cycle* (*path*) is a spanning subgraph that is a cycle (path). A *ray* in an infinite graph is an infinite one-way path; that is, a path with vertices in the set $\mathbb{N}$, and edges $i(i+1)$ for all $i \in \mathbb{N}$. A *Hamilton path* in an infinite graph is a spanning subgraph that is a ray.

The *complement* of G, written $\overline{G}$, has vertices $V(G)$ with two distinct vertices joined if and only if they are not joined in G. A *complete graph of order n* or *n-clique* has all edges present, and is written K_n. A set of vertices S is *independent* or *stable* if $\langle S \rangle_G$ contains no edges. The graph of order n with no edges is $\overline{K_n}$. A *wheel of order n*, written W_n, consists of a cycle C_n along with one universal vertex. A *hypercube* Q_n has vertices elements of $\{0, 1\}^n$ (that is, all binary sequences of length n), with two vertices joined if they differ in exactly one coordinate.

The *chromatic number* of G, written $\chi(G)$, is the minimum cardinal n with the property that $V(G)$ may be partitioned into n independent sets; that is, the minimum n so that G has a *proper n-colouring.* If $\chi(G) = 2$, then G is *bipartite.* The $\chi(G)$-many independent sets are sometimes called *vertex classes* or *colours.* We may think of $\chi(G)$ as the minimum number of colours needed so each vertex is coloured with a unique colour, and any edge has endpoints with distinct colours. It is an exercise to check that $\chi(G) \leq n$ if and only if there is some homomorphism from G to K_n. For this reason, homomorphisms are often thought of as generalizations of proper n-colourings. The bipartite graphs are characterized as follows.

THEOREM 1.5. *A graph G is bipartite if and only if it has no odd cycle as a subgraph.*

Proof. The forward direction follows immediately, since a subgraph of a bipartite G is bipartite. For the reverse direction, without loss of generality suppose that G is connected. Fix u in $V(G)$, and let A be the set of vertices of even distance to u, and B is the set of vertices of odd distance to u. Then A and B partition $V(G)$. An edge between vertices of A or between vertices of B creates an odd cycle. □

A *complete bipartite graph* has all possible edges present between the two colours. These graphs are denoted by $K_{m,n}$, where m and n are the orders of the vertex classes. A *star* is a graph $K_{1,n}$, for some positive integer n.

The *clique number* of G, written $\omega(G)$, is the order of a largest clique (more precisely, the supremum of all orders of cliques), while the *independence* (or *stability*) number of G, written $\alpha(G)$, is the order of a largest independent set. See Figure 1.3 for an example illustrating these graph parameters. The following elementary inequalities are useful (see Exercise 6).

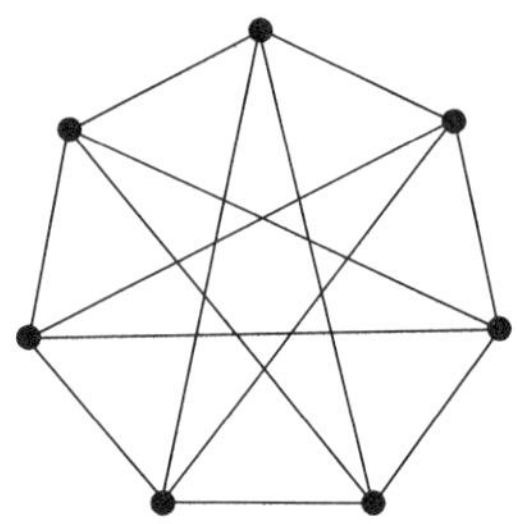

Figure 1.3. A graph G with $\omega(G) = \alpha(G) = 3$ and $\chi(G) = 4$.

THEOREM 1.6. *If G is a graph, then $\omega(G) \leq \chi(G)$, while*

$$\chi(G) \geq |V(G)|/\alpha(G).$$

In a graph G, a set S of vertices is a *dominating set* if every vertex not in S has a neighbour in S. The *domination number* of G, written $\gamma(G)$, is the minimum cardinality of a dominating set.

A *directed graph* or *digraph* is defined identically as a graph, except that $E(G)$ need not be a symmetric binary relation on $V(G)$. The edges are written as ordered pairs, and are called *directed edges* or *arcs* (u, v), where u is the *head* and v is the *tail*. The vertex v is an *out-neighbour* of u, while u is an *in-neighbour* of v. The *in-degree* of u, written $\deg^{-}(u)$, is the number of vertices v such that (v, u) are directed edges; the *out-degree* $\deg^{+}(u)$ is defined dually.

Subgraphs, induced subgraphs, and isomorphism are defined analogously to those for graphs. Paths and cycles are *directed* if there are no pairs of edges (x, y), (z, y) for distinct vertices x, y, and z. See Figure 1.4.

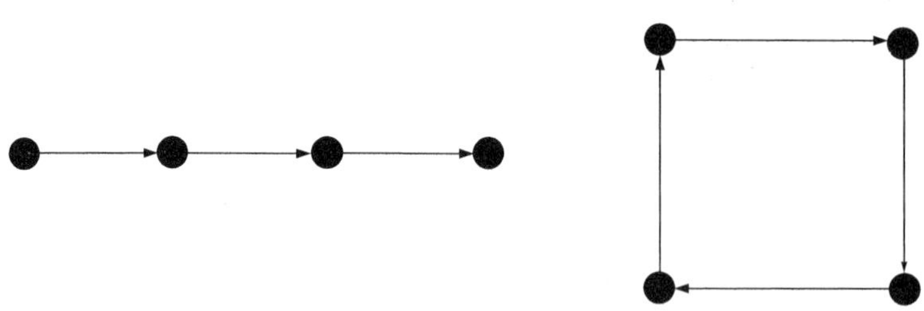

Figure 1.4. A directed path and cycle.

For a digraph G, define the equivalence relation R so that xRy if and only if there is a directed path from x to y and a directed path from y to x. The equivalence classes are called *strongly connected components* of G, and a graph is strongly connected if there is only one strongly connected component. We may assign a direction to each edge of a graph G to obtain a digraph called an *orientation* of G. An orientation of a complete graph is called a *tournament*. An *acyclic orientation* of a graph is one without directed cycles.

1.3. Probability Theory

As any mathematician can tell you, there is more to probability than gambling. Indeed, applications of probability theory are now common in both graph theory and theoretical computer science. As we will witness in Chapters 3 and 4, a grasp of expectation, variance, and concentration are essential prerequisites to the study of models for the web graph.

For simplicity and utility, we will restrict ourselves to discrete probability spaces. A pleasant side effect of this restriction is that we do not require background in measure theory. A *(discrete) probability space* $\mathcal{S}$ consists of a triple $(S, \mathcal{F}, \mathbb{P})$. The set S, called the *sample space*, is non-empty and countable (although we usually take S to be finite). The set $\mathcal{F}$ is the collection of all subsets of S; the elements of $\mathcal{F}$ are *events*. The function $\mathbb{P} : \mathcal{F} \rightarrow \mathbb{R}$, named the *probability measure*, satisfies the following properties.

(1) For all events A, $\mathbb{P}(A) \in [0, 1]$, and $\mathbb{P}(S) = 1$.

(2) If $(A_i : i \in I)$ is a countable set of events that are pairwise disjoint, then

$$\mathbb{P}\left(\bigcup_{i \in I} A_i\right) = \sum_{i \in I} \mathbb{P}(A_i).$$

Since we always take $\mathcal{F}$ to be the power set of S, we need not specify $\mathcal{F}$ when defining a discrete probability space. Given any finite set S with $n > 0$ elements, as a good first example of a probability space we may consider the *uniform probability space* on S, with

$$\mathbb{P}(A) = \frac{|A|}{n}.$$

The reader may verify that items 1 and 2 in the definition of a probability space hold. Observe that the probability of choosing any fixed element of S is $\frac{1}{n}$. An element chosen with probability $\frac{1}{n}$ from S is said to be chosen *uniformly at random*, also written *u.a.r.*

The proof of the following lemma follows from the definition of a probability space.

LEMMA 1.7. *If $(S, \mathcal{F}, \mathbb{P})$ is a probability space and $A, B \in \mathcal{F}$, then*

(1) $\mathbb{P}(\varnothing) = 0$.

(2) $\mathbb{P}(S \backslash A) = 1 - \mathbb{P}(A)$.

(3) *If* $A \subseteq B$, *then* $\mathbb{P}(A) \le \mathbb{P}(B)$.

(4) *If* $(A_i : i \in I)$ *is a countable set of events, then*

$$\mathbb{P}\left(\bigcup_{i\in I} A_i\right) \le \sum_{i\in I} \mathbb{P}(A_i).$$

Events A and B are *independent* if $\mathbb{P}(A \cap B) = \mathbb{P}(A)\mathbb{P}(B)$, while a set $(A_i : 1 \le i \le n)$ of events is *mutually independent* if

$$\mathbb{P}\left(\bigcap_{i=1}^{n} A_i\right) = \prod_{i=1}^{n} \mathbb{P}(A_i).$$

If $\mathbb{P}(B) > 0$, then the *conditional probability* that A occurs given B is

$$\mathbb{P}(A|B) = \frac{\mathbb{P}(A \cap B)}{\mathbb{P}(B)}.$$

A (*discrete*) *random variable* X on a probability space $\mathcal{S}$ is a function $X : S \to \mathbb{R}$. For example, given a fixed real number $p \in (0, 1]$, let $S = \mathbb{N}\backslash\{0\}$ and define

$$\mathbb{P}(x) = p(1-p)^{x-1}.$$

(You should convince yourself that this is a probability measure.) Define $X(x) = x$. The random variable X is a *geometric random variable with parameter* p. A geometric random variable represents the number of failed trials required to obtain the first success; for example, it represents the number of tosses of a coin until the first head appears.

Given $x \in \mathbb{R}$, define

$$\mathbb{P}(X = x) = \sum_{\substack{s\in S,\\ X(s)=x}} \mathbb{P}(\{s\}).$$

The probabilities $\mathbb{P}(X \ge x)$, $\mathbb{P}(X > x)$, $\mathbb{P}(X \le x)$, and $\mathbb{P}(X < x)$ are defined analogously. The function $f : \mathbb{R} \to [0, 1]$ defined by $f(x) = \mathbb{P}(X = x)$ is called the *probability mass function of* X. The expectation (also called the *mean, average,* or *first moment*) of a random variable X, written $\mathbb{E}(X)$, is defined by

$$\mathbb{E}(X) = \sum_{s\in S} X(s)\mathbb{P}(\{s\}).$$

If S is finite, then $\mathbb{E}(X)$ is always finite; otherwise, $\mathbb{E}(X)$ may be infinite. If $X \ge 0$ (which is usually the case for the random variables we consider), then $\mathbb{E}(X) \ge 0$. For the example of a geometric random variable with parameter

p mentioned earlier, we have that

$$\begin{aligned}\mathbb{E}(X) &= \sum_{x\in\mathbb{N}\setminus\{0\}} xp(1-p)^{x-1}\\ &= \frac{p}{(1-(1-p))^2}\\ &= \frac{1}{p}.\end{aligned}$$

Expectation has the following properties, each of which follows from the definitions.

THEOREM 1.8. *Suppose that X, Y, and X_i, where $1 \le i \le n$, are random variables defined on a probability space. Then the following properties hold.*

(1) (*Linearity of expectation*) *Let c_i, where $1 \le i \le n$, be real numbers. Then*

$$\mathbb{E}\left(\sum_{i=1}^{n} c_i X_i\right) = \sum_{i=1}^{n} c_i \mathbb{E}\left(X_i\right).$$

(2) (*Monotonicity*) *If $X \le Y$ (that is, $X(s) \le Y(s)$ for all $s \in S$), then $\mathbb{E}(X) \le \mathbb{E}(Y)$.*

Random variables X and Y are *independent* if for all real numbers x and y the events that $X \le x$ and $Y \le y$ are independent. The *variance* of a random variable X, written $\mathrm{Var}(X)$, is defined by

$$\begin{aligned}\mathrm{Var}(X) &= \mathbb{E}((X-\mathbb{E}(X))^2)\\ &= \mathbb{E}(X^2) - \mathbb{E}(X)^2,\end{aligned}$$

where the second equality follows by the linearity of expectation. The quantity $\mathrm{Var}(X)$ is a measure of how far X deviates from $E(X)$. Note that $\mathrm{Var}(X) \ge 0$. The *covariance* of two random variables X and Y defined on the same probability space is

$$\mathrm{Cov}(X,Y) = \mathbb{E}((X-\mathbb{E}(X))(Y-\mathbb{E}(Y))).$$

Two basic but very useful inequalities on random variables are contained in the following theorems.

THEOREM 1.9 (Markov's inequality). *Let $X \ge 0$ be a random variable on a probability space. If c is a positive real number, then*

$$\mathbb{P}(X \ge c) \le \frac{\mathbb{E}(X)}{c}.$$

Proof. The proof is clear from the following inequalities:

$$\begin{aligned}\mathbb{E}(X) &= \sum_{s\in S}\mathbb{P}(\{s\})X(s)\\ &\geq \sum_{\substack{s\in S,\\ X(s)\geq c}}\mathbb{P}(\{s\})X(s)\\ &\geq \sum_{\substack{s\in S,\\ X(s)\geq c}}\mathbb{P}(\{s\})c\\ &= c\mathbb{P}(X\geq c).\end{aligned}$$ □

THEOREM 1.10 (Chebyshev's inequality). *Let X be a random variable on a probability space. If c is a positive real number, then*

$$\mathbb{P}(|X-\mathbb{E}(X)|\geq c)\leq \frac{\mathrm{Var}(X)}{c^2}.$$

Proof. By Theorem 1.9, we have that

$$\mathbb{P}(|X-\mathbb{E}(X)|\geq c)=\mathbb{P}((X-\mathbb{E}(X))^2\geq c^2)\leq \frac{\mathrm{Var}(X)}{c^2}.$$ □

We now consider some well-known examples of random variables. Suppose we run an experiment so that the probability of success is p, while the probability of failure is $1-p$. For example, if we toss a fair coin, then the probability of a head or tail is $p=1-p=\frac{1}{2}$. If a random variable X takes values 1 and 0 with probability p and $1-p$, respectively, then we say the X is a *Bernoulli* or *indicator* random variable. Then $\mathbb{E}(X)=p$. An easy calculation gives the variance of X:

$$\begin{aligned}\mathrm{Var}(X) &= \mathbb{E}(X^2)-\mathbb{E}(X)^2\\ &= p-p^2\\ &= p(1-p).\end{aligned}$$

If we form the sum of n independent and identically distributed Bernoulli random variables, then we obtain a binomial random variable. More precisely, a *binomial random variable* X with parameters n and p, written $X\in Bi(n,p)$, has the probability mass function

$$\mathbb{P}(X=i)=\binom{n}{i}p^i(1-p)^{n-i}.$$

For example, if we toss a fair coin n times and X counts the number of heads, then $X\in Bi(n,\frac{1}{2})$ and the probability of obtaining i heads equals

$$\mathbb{P}(X=i)=\binom{n}{i}\left(\frac{1}{2}\right)^n.$$

If $X \in Bi(n,p)$, then $\mathbb{E}(X) = np$ and the variance equals $np(1-p)$. Clearly, if $n = 1$, then $X \in Bi(n,p)$ is a Bernoulli random variable.

Let X be a random variable which takes values on $\mathbb{N}$. Then X is a *Poisson random variable* if it has the probability mass function

$$\mathbb{P}(X = i) = \frac{\lambda^i}{i!}e^{-\lambda},$$

for some $\lambda > 0$. Consider tossing m balls into n bins, with the location of each ball chosen independently and u.a.r. from the n bins. If X is the number of balls in a bin, then X is approximately Poisson with $\lambda = m/n$ (which is the average number of balls per bin).

If X is Poisson, then $\mathbb{E}(X) = \mathrm{Var}(X) = \lambda$. The following theorem demonstrates that if n is large and p is small, then in certain cases the Poisson distribution is the limiting distribution of the binomial distribution (see Section 3.5 of [**122**] for further discussion and a proof).

THEOREM 1.11. *Let $(X_n : n \in \mathbb{N})$ be a sequence of random variables with each $X_n \in Bi(n,p)$, where $\lim_{n\to\infty} np = \lambda > 0$, and λ is independent of n. Then for all $i \in \mathbb{N}$,*

$$\lim_{n\to\infty} \mathbb{P}(X_n = i) = \frac{\lambda^i}{i!}e^{-\lambda}.$$

The Markov and Chebysev inequalities apply to any random variable X and measure deviation from the expected value of X. However, the bounds supplied by these inequalities are not always tight. For a binomial random variable, the probability of deviating from the mean tends to 0 exponentially the larger the distance to the expectation. This is made precise by the following set of inequalities, called *Chernoff bounds.* For a proof, see Theorem 2.1 and Corollary 2.3 of [**137**].

THEOREM 1.12. *Let X be a binomial random variable $X \in Bi(n,p)$ with $\mathbb{E}(X) = np$. The following bounds hold:*

$$\begin{aligned}\mathbb{P}(X \geq \mathbb{E}(X) + t) &\leq \exp\left(-\mathbb{E}(X)\,\phi(t/\mathbb{E}(X))\right),\\ \mathbb{P}(X \leq \mathbb{E}(X) - t) &\leq \exp\left(-\mathbb{E}(X)\,\phi(-t/\mathbb{E}(X))\right),\end{aligned}$$

where $\phi(x) = (1+x)\log(1+x) - x$ for $x \geq -1$, and $\phi(x) = \infty$ for $x < -1$. If $\varepsilon \leq 3/2$, then

$$\mathbb{P}\left(|X - \mathbb{E}(X)| \geq \varepsilon\mathbb{E}(X)\right) \leq 2\exp\left(-\frac{1}{3}\varepsilon^2\mathbb{E}(X)\right).$$

A *random* (or *stochastic*) *process* on a probability space is a set $(X_i : i \in I)$ of countably many random variables on the space. An important example of a random process is a Markov process, where the future is conditionally independent of the past given the present. Markov processes will be introduced formally in Chapter 5. Other random processes modelling the evolution of W will be presented in Chapter 4.

Exercises

(1) Suppose that $0 \leq m \leq n$ are integers with $n > 0$.
  (a) Show that
$$\binom{n}{m} \leq \frac{n^m}{m!} \leq n^m.$$
  (b) Prove that
$$\left(\frac{n}{m}\right)^m \leq \binom{n}{m} \leq \left(\frac{ne}{m}\right)^m.$$

(2) (a) If a graph G contains a cycle, then $g(G) \leq 2\,\mathrm{diam}(G) + 1$.
  (b) Prove that for all graphs G, either G or $\overline{G}$ is connected.
  (c) Prove that for all graphs G, some two vertices have the same degree.

(3) (a) Prove that a tree T has at least $\Delta(T)$ vertices of degree 1 (also called *endvertices* or *leaves*).
  (b) The *eccentricity* of a vertex u in a graph G is the maximum of all the distances $d(u, v)$, where v is a vertex of G. The *centre* of a graph is the subgraph induced by the vertices of minimum eccentricity. Prove that the centre of a tree is a vertex or an edge.

(4) (a) Prove that for a graph G, $\chi(G) \leq \Delta(G) + 1$.
  (b) Prove *Brook's theorem*: if G is connected and is neither a complete graph nor an odd cycle, then $\chi(G) \leq \Delta(G)$.

(5) (a) Show that if G is an infinite and connected graph, then some vertex has infinite degree or G contains a ray as a subgraph.
  (b) Show that a connected uncountable graph contains some vertex of uncountable degree.
  (c) Supply a family of $2^{\aleph_0}$-many non-isomorphic countable trees where each vertex has finite degree.

(6) Prove Theorem 1.4.

(7) Prove Theorem 1.6.

(8) (a) Show that a graph G admits a homomorphism to K_n if and only if $\chi(G) \leq n$.
  (b) Prove that for all $n \geq 1$, C_{2n+3} admits a homomorphism to C_{2n+1}, but the converse is false.

(9) A graph is a *core* if it does not admit a homomorphism to a proper subgraph.
   (a) Prove that cliques, odd cycles, odd wheels, and the Petersen graph are cores.
   (b) A graph is *vertex-critical* if for all $x \in V(G)$,
$$\chi(G - x) < \chi(G).$$
   Show that every vertex-critical graph is a core.
   (c) Show that if G is a vertex-transitive graph such that
$$\gcd(|V(G)|, \alpha(G)) = 1$$
   (where $\gcd(m, n)$ is the greatest common divisor of integers m and n), then G is a core.
   (d) Let m, n be integers such that $1 \le m < \frac{n}{2}$. The *Kneser graph* $K(m, n)$ has vertices that are the m-subsets of $[n]$, and two vertices are joined if and only if they are disjoint. Show that all Kneser graphs are cores. By using tools from algebraic topology, Lovász [**156**] proved that $\chi(K(m, n)) = n-2m+2$ (a purely combinatorial proof was later given in [**159**]).

(10) Prove that the hypercubes Q_n, where $n \ge 2$, are bipartite, vertex-transitive, and have a Hamilton cycle.

(11) A graph G is *self-complementary* if and only if $\overline{G} \cong G$. Prove that a self-complementary graph G exists if and only if one of $|V(G)|$ or $|V(G)| - 1$ is divisible by 4.

(12) (a) Determine the automorphism group of the following graphs: K_n, C_n, P_n, the Petersen graph.
   (b) A graph is *asymmetric* (or *rigid*) if it has no non-trivial automorphism. Construct infinitely many asymmetric graphs.

(13) (a) Show that for a digraph G,
$$\sum_{u \in V(G)} \deg^-(u) = \sum_{u \in V(G)} \deg^+(u) = |E(G)|.$$
   (b) A *king* in a tournament is a vertex from which every vertex is reachable by a path of length at most two. Prove that every tournament has a king.
   (c) A *topological sort* of a digraph is a numbering of the vertices such that every edge is increasing (that is, goes from smaller to greater number). Show that a digraph G has a topological sort if and only if G has no directed cycles.

(14) Prove Lemma 1.7 and Theorem 1.8.

(15) Show that if $X \in Bi(n, p)$, then $\mathbb{E}(X) = np$ and $\mathrm{Var}(X) = np(1 - p)$.

(16) (a) Suppose that X and Y are independent random variables. Prove the following equalities.
i) $\mathbb{E}(XY) = \mathbb{E}(X)\mathbb{E}(Y)$.
ii) $\mathrm{Var}(X + Y) = \mathrm{Var}(X) + \mathrm{Var}(Y)$.
iii) $\mathrm{Cov}(X, Y) = 0$.
(b) Give counterexamples to the equalities in i), ii), and iii) if X and Y are not independent.

(17) A function $f : \mathbb{R} \to \mathbb{R}$ is *convex* if for all $x, y \in \mathbb{R}$ and $0 \leq c \leq 1$,
$$f(cx + (1 - c)y) \leq cf(x) + (1 - x)f(y).$$
Prove *Jensen's inequality*: If f is a convex function, then
$$f(\mathbb{E}(X)) \leq \mathbb{E}(f(X)).$$

(18) Let $(X_i : 1 \leq i \leq n)$ be independent and identically distributed random variables, each with expected value u and variance v. Define
$$\overline{X_n} = \frac{1}{n}\sum_{i=1}^{n} X_i.$$
(a) Prove that $\mathbb{E}(\overline{X_n}) = u$ and $\mathrm{Var}(\overline{X_n}) = \frac{v}{n}$.
(b) Using Chebyshev's inequality and (a), prove the *weak law of large numbers*: for every real $\varepsilon > 0$,
$$\lim_{n \to \infty} \mathbb{P}\left(\left|\overline{X_n} - u\right| \leq \varepsilon\right) = 1.$$

(19) Let $X \in Bi(n, \frac{1}{2})$. Using the Chernoff bounds, show that
$$\mathbb{P}\left(\left|X - \frac{n}{2}\right| \geq \frac{1}{2}\sqrt{6n \log n}\right) \leq \frac{2}{n}.$$

(20) Suppose that a coin lands heads with probability p and tails with probability $1 - p$. Let E_n be the event that in a sequence of $n > 0$ coin tosses, an even number of heads appear.
(a) Prove that for all $n \in \mathbb{N} \setminus \{0\}$,
$$\mathbb{P}(E_n) = p(1 - \mathbb{P}(E_{n-1})) + (1 - p)\mathbb{P}(E_{n-1}).$$
(b) Using the recurrence relation of part (a), prove that
$$\mathbb{P}(E_n) = \frac{1}{2}(1 + (1 - 2p)^n).$$

(21) Assign n balls to n bins, where each ball is assigned independently and u.a.r.
(a) Let X be the number of balls assigned before all bins are non-empty. Show that $\mathbb{E}(X) = O(n \log n)$, and $\mathrm{Var}(X) = O(n^2)$.

(b) Let Y be the number of non-empty bins. Prove that

$$\mathbb{E}(Y) = n\left(1 - \left(1 - \frac{1}{n}\right)^m\right).$$

(22) Given a finite set of probability spaces

$$\{(S_i, \mathcal{F}_i, \mathbb{P}_i) : 1 \leq i \leq m\}$$

we may form the *product space* $(S, \mathcal{F}, \mathbb{P})$ with sample space $\prod_{i=1}^m S_i$, events $\prod_{i=1}^m \mathcal{F}_i$, and for an event $A = \prod_{i=1}^m A_i$,

$$\mathbb{P}(A) = \prod_{i=1}^m \mathbb{P}_i(A_i).$$

(a) Prove that the product space is a probability space.
(b) Fix n a positive integer, $p \in [0, 1]$, and $i \in [n]^2$. Let $S_i = \{0_i, 1_i\}$, $\mathbb{P}_i(\{1_i\}) = p$, and $\mathbb{P}_i(\{0_i\}) = 1 - p$. Verify that for a fixed i this gives rise to a probability space $(S_i, \mathcal{F}_i, \mathbb{P}_i)$. The corresponding product space over all $i \in [n]^2$, written $G(n, p)$, is called *a random graph with n vertices and edge probability p*.
(c) For $i \in [n]^2$, define E_i to be the event consisting of the set of $s \in S$ whose ith coordinate is 1_i. Prove that the events $\{E_i : i \in [n]^2\}$ are mutually independent, and $\mathbb{P}(E_i) = p$. We may therefore identify the probability space $G(n, p)$ with graphs whose vertex set is $[n]$, and whose edges are chosen independently and with probability p.

Chapter 2

The Web Graph

If all art aspires to the condition of music, then all the sciences aspire to the condition of mathematics.

–George Santayana

The latest authors, like the most ancient, strove to subordinate the phenomena of nature to the laws of mathematics.

–Isaac Newton

2.1. Introduction

Our primary focus is on the mathematics surrounding a certain real-world network called the *web graph*, which we will denote by W. The graph W has vertices representing web pages, and edges representing the links between pages. Unlike the abstract graphs discussed in Chapter 1, W is an experimental, real-world graph. Speaking of *the* web graph is somewhat misleading, as the web is an evolving structure with pages and links appearing and disappearing continuously over time. We will overlook such concerns for the remainder of the book, and treat W as any other graph.

What are the key properties of W? As we will learn, the massive, evolving graph W has the following properties: it is a *sparse, self-organizing, small world, and power law network.* All these properties will be defined in this chapter. Along the way, we will learn that W is not the only real-world graph with these properties. Many technological, social, biological networks have properties similar to those present in the web. For example, power laws have been observed in protein-protein interaction networks, and networks formed by scientific collaborators.

In this first section, we summarize the experimental data observed in W collected from various web crawls. The main reason for presenting this information is to supply some context for the modelling of W, described later in Chapter 4. We focus on the following properties of W: its order (Section 2.1.1), power law degree distribution (Section 2.1.2), the small world property (Section 2.1.3), and community structure (Section 2.1.4). For additional discussion of empirical results of other properties studied in W (such as orders of connected components and betweenness) see [**88**]. In Section 2.2, we survey other real-world networks distinct from W which share some of its properties.

The terms "world wide web" and "internet" are often treated as synonymous, but they actually refer to different things. The *internet* is the more general term, including various physical components and hardware, as well as other aspects such as web and e-mail. As we will describe in Section 2.2, the internet may be viewed on various levels, such as the interdomain and routing level. Broadly speaking, the internet consists of a vast network of smaller interconnected networks, exchanging data by certain protocols. The birth of the internet is thought to be in the early 1980's, when the National Science Foundation constructed a network connecting colleges and universities. In the coming years, the internet became increasingly open to commercial interests, and became part of the public's consciousness by the mid 1990's. For a time-line of the history of the internet, ranging from the 1950's to the present, the reader is directed to [**197**].

The *world wide web* consists of information stored and available on the internet. Most web pages are *Hypertext Markup Language* (*HTML*) documents identified with strings called *Uniform Resource Locators* (*URLs*). HTML documents are joined by *links* (or *hyperlinks*), which may be taken as directed or undirected edges in W, depending on context.

A discussion of the technologies underlying the internet and the web is beyond the scope of this book. For an introduction to these technologies, the reader is directed to Chapter 2 of [**19**].

2.1.1. How big is the web? Even a casual surf on the web will convince you that there are an enormous number of web pages and links. In the mid-1990's several search engines claimed to cover the entire web. In 1998, Lawrence and Giles [**150**] claimed that there were at least 320 million web pages in late 1997. A 1999 study [**151**] by the same authors uncovered around 800 million pages. The web was evidently growing very quickly around the turn of the millennium.

As W is growing rapidly, it seems difficult to obtain an exact, current estimate of the number of web pages. In 2005, [**125**] reported that the web graph had about 11.5 billion pages. A recent study of Hirate et al. [**132**] found 53.7 billion web pages, with 34.7 billion pages indexed by *Google*.

The web is often cited as possessing a *bow-tie* structure, first described in [**54**]. The knot of the bow-tie consists of a strongly connected component, called the *core* or *SCC*. One side of the bow consists of pages which link to the core, with the other side consisting of pages which contain a link from the core. The remaining pages are either in disconnected components, or in *tendrils* that only link to pages outside the core. Estimates have been given for the order of the various regions. For example, while it was estimated in [**54**] that around one third of pages were contained in the core, a recent estimate of [**132**] places more than two thirds of all web pages in the core. The apparent growth in the core over time reflects the increasing connectivity between web pages.

Another view of W is as an *infinite* graph. To explain this remark, a *static* web page provides the same downloadable information to all users, while the remaining pages are *dynamic*. For example, my website is static, while an on-line calendar is dynamic. The number of dynamic web pages is infinite (although practically finite owing to memory and other resource constraints). See Chapter 6 for more on this view of W. We therefore assume the following caveat.

For all the chapters except Chapter 6, we will only consider finite graphs.

Regardless of its exact order, the web is a massive graph. The *average degree* of a graph G is defined as

$$\frac{1}{|V(G)|} \sum_{v \in V(G)} \deg_G(v) = \frac{2|E(G)|}{|V(G)|}.$$

We say that a graph G is *sparse* if the average degree of G is at most $\epsilon|V(G)|$, where $\epsilon < 1$. Otherwise, we say that G is *dense*. See Figure 2.1.

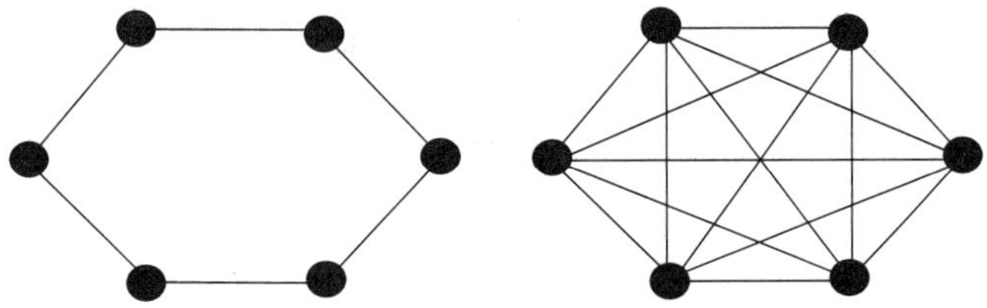

Figure 2.1. A sparse and a dense graph.

We note that an alternative definition of sparse, which does not apply to W, is that $|E| = O(|V(G)|)$. We should expect W to be quite sparse, since the average number of links from any one fixed web page is bounded above by a constant (see Chapter 3 of [**19**]). It is worth noting that W

behaves quite differently when viewed as a directed or undirected graph. For example, an author of a web page p has no (or limited) control over the in-degree of p, but has usually complete control over the out-degree of p.

2.1.2. Power law degree distributions. Arguably the most important properties observed in W are power law degree distributions. Given an undirected graph G and a non-negative integer k, we define $N_{k,G}$ by

$$N_{k,G} = |\{x \in V(G) : \deg_G(x) = k\}|.$$

The parameter $N_{k,G}$ is the *number of vertices of degree k in G*. For simplicity, suppose that $|V(G)| = t$. Then $N_{k,G}$ is an integer in the interval $[0, t]$. We will suppress the subscript G if it is clear from the context. The *degree distribution* of G is the sequence $(N_{k,G} : 0 \leq k \leq t)$. See Figure 2.2 for an example.

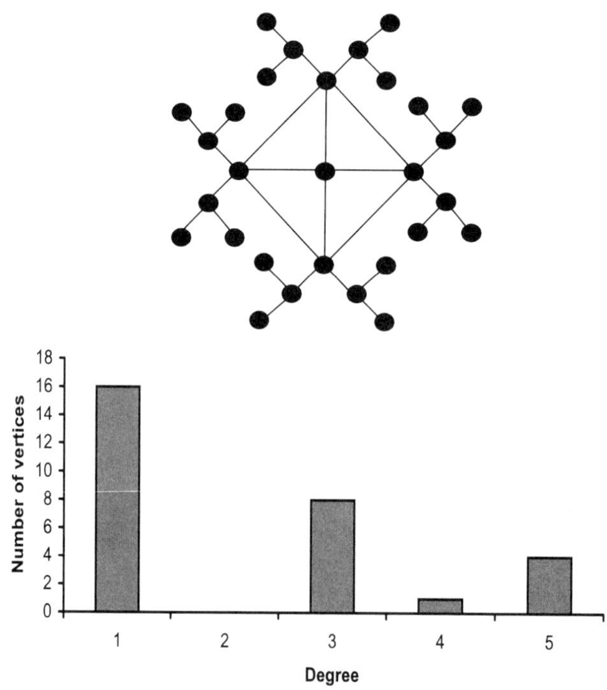

Figure 2.2. The graph G and a plot of its degree distribution.

We say that the degree distribution of G follows a *power law* if for each degree k,

$$\frac{N_{k,G}}{t} \sim k^{-\beta}, \tag{2.1}$$

for a fixed real constant $\beta > 1$. Note that (2.1) is asymptotic, as we will often use it in application to an infinite family of finite graphs generated by a random graph model (we may interpret (2.1) for a fixed graph as meaning that $\frac{N_{k,G}}{t}$ is approximately $k^{-\beta}$). The reason for this is that since W is a massive graph, we are more interested in the approximate rather than exact value of $\frac{N_{k,G}}{t}$.

Power law distributions are sometimes called *heavy-tailed distributions*, since the real-valued function $f(k) = k^{-\beta}$ exhibits a polynomial (rather than exponential) decay to 0 as k tends to ∞. We say that β is the *exponent of the power law*. If G possesses a power law degree distribution, then we simply say G is a *power law graph*. See Figure 2.3 for an instance of a power law graph. A large number of low degree vertices are drawn on the periphery,

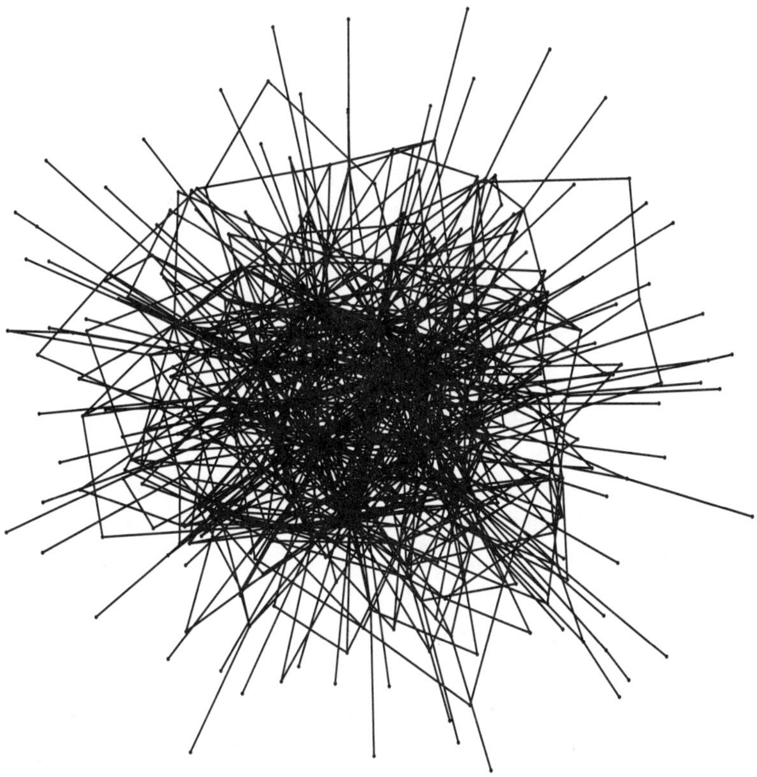

Figure 2.3. A power law graph with 400 vertices.

with the high degree vertices clustered toward the centre of the figure. If we take logarithms on both sides of (2.1), then the relationship is expressed as

$$\log(N_{k,G}) \sim \log(t) - \beta \log(k).$$

Hence, in the log-log plot, we obtain a straight line with slope $-\beta$.

In both real-world networks and graphs generated by theoretical models, the power law may only fit for a certain range of degrees, with discrepancies for small or large degree vertices. For an example, see the log-log plot of the degree distribution of a power law graph in Figure 2.4, where a straight line is quite evident for lower degrees. However, there is more scattering among the higher degrees.

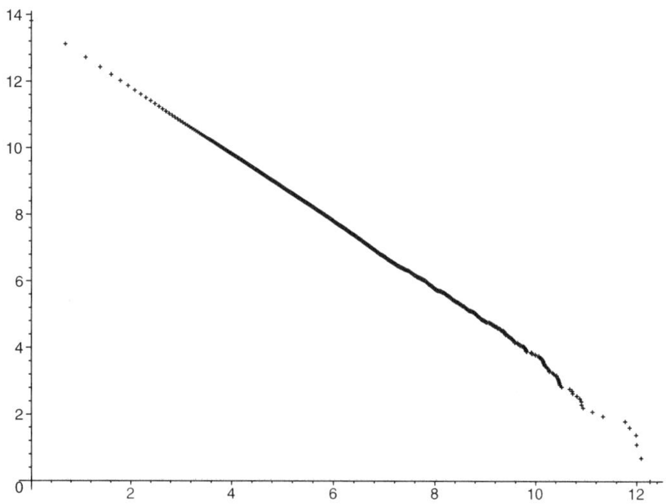

Figure 2.4. The log-log plot of the degree distribution of a power law graph.

The graph W may clearly be viewed as either a directed or an undirected graph. If G is directed, then we may discuss power laws for the in- and out-degree distributions by defining $N^{in}_{k,G}$ and $N^{out}_{k,G}$, respectively, in the obvious way.

Power laws have a long history, stretching back to Pareto's work [**171**] in 1896 on income distribution. For a thorough discussion of this history, the reader is directed to [**163**]. The empirical study of W began before the turn of the millennium. Based on their crawl of the domain of Notre Dame University, Albert, Barabási, and Jeong [**8**] claimed that the web graph exhibited a power law in-degree distribution, with $\beta = 2.1$. An independent crawl corroborating the findings of [**8**] was reported from IBM researchers [**144**], who considered 40 million web pages from 1997 data. The exponent of $\beta = 2.1$ was further corroborated by a larger crawl of the entire web (including 200 million web pages) reported in Broder et al. [**54**]. There was some evidence presented in both studies that the out-degree distribution follows a power law with [**8**] reporting $\beta = 2.45$ and [**54**] reporting $\beta = 2.72$.

The presence of power law degree distributions reflects a certain *undemocratic* aspect of W: while most pages have a small number of links, a few pages have a large number. This is in hindsight not surprising, since the choice of links from new pages to existing ones is determined by the users' own interests. For example, it seems plausible that popular pages attract more new links than unpopular ones. Possible mechanisms underlying the power law degree distribution of W will be presented in Chapter 4. Generating graphs with power law degree distributions has become a hot topic for random graph theorists, leading to a growing corpus of rigorous results that will be further discussed in Chapter 4.

While our focus is on properties of W, it is worthwhile noting that power laws (and several other properties of W) are indicative of choice in many other contexts. Power laws are now known to be pervasive in a variety of real-world man-made networks where some degree of choice is involved, such as the telephone call network, the e-mail network, or the scientific citation network. Power law degree distributions are also prevalent in social networks (such as friendship networks), and biological networks (such as the network of protein-protein interactions in a cell). In biological networks, evolution is the dominant decision making force in the generation of vertices and edges. See Section 2.2 for a more detailed discussion of other real-world power law graphs.

2.1.3. The small world property. With technologies available such as the e-mail and cell-phones, the world definitely feels like a smaller place. The term *small world graphs* was first introduced by social scientists Watts and Strogatz [**194**] in their study of various real-world networks, such as the network of Hollywood movie actors. The paper [**194**] introduced the average distance (or characteristic path length) which measures global distances in a graph, and the clustering coefficient, which is a measure of "cliquishness" of neighbourhoods in a graph. We now define both of these concepts.

The diameter of a graph is a well-known global measure of distances in a graph. Small world graphs G of order t should satisfy

$$\mathrm{diam}(G) = \Theta(\log t).$$

Despite this condition, data from [**54**] suggest that $\mathrm{diam}(W) > 900$. In a real sense the diameter of W is infinite: simply create a web page p with no links, and ensure that no one knows about it. Then p is an isolated vertex in W!

Another measure of distance, introduced in [**194**], is defined by

$$L(G) = \sum_{u,v \in S} \frac{d(u,v)}{|S|},$$

where S is the set of pairs of distinct vertices u, v of G with the property that $d(u,v)$ is finite. If G is connected with order t, then this becomes

$$\begin{aligned} L(G) &= \sum_{u,v\in V(G)} d(u,v) \binom{t}{2}^{-1} \\ &= \frac{2}{t^2-t} \sum_{u,v\in V(G)} d(u,v). \end{aligned}$$

(For a connected graph, the graph parameter $\sum_{u,v\in V(G)} d(u,v)$ is sometimes called the *Wiener index*; see [**195**].) The rational number $L(G)$ is the *average distance* of G. See Figure 2.5. The directed analogue of this parameter, where distance refers to shortest *directed* paths (that is, with no back edges), is denoted $L_d(G)$. The definition of $L(G)$ avoids the above-mentioned problems with diameter and isolated vertices, and focusses on distances between vertices in the same component. The small world property [**194**] demands that $L(G)$ (or $L_d(G)$ if G is directed) must be much smaller than the order of the graph. To make this precise, we may insist that

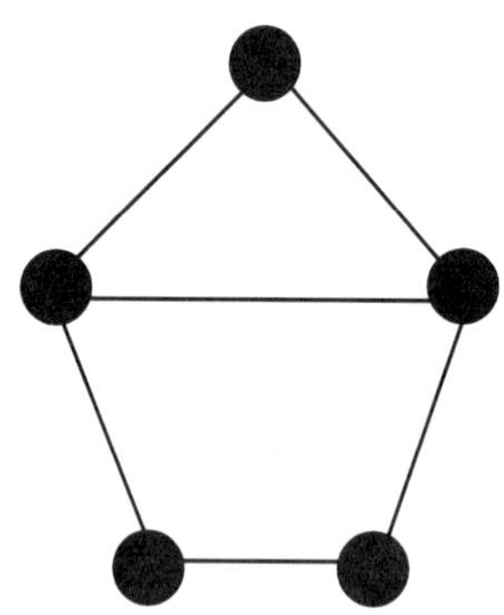

Figure 2.5. A graph G with $L(G) = 1.4$.

$$L(G) = \Theta(\log\log t)$$

(as is the case for certain random power law graphs; see Theorem 7.7 of [**78**] or Theorem 4.21 in Chapter 4). As evidence of the small world property for W, in [**8**] it was reported that $L_d(W) = 19$, while [**54**] reported $L_d(W) = 16$ and $L(W) = 6.8$.

We now turn to the clustering coefficient of a graph, which measures local density. According to [**194**], while we would expect $L(G)$ to be low in a small world graph, the clustering coefficient should be "large": they specify this to mean that the clustering coefficient should be larger than the clustering coefficient of a random graph with the same number of vertices and edges. (See Chapter 3 for more on random graphs.) For a graph G of order t and $x \in V(G)$, define

$$\begin{aligned} C(x) &= \frac{|E(G \restriction N(x))|}{\binom{\deg(x)}{2}} \\ &= \frac{2|E(G \restriction N(x))|}{\deg(x)(\deg(x)-1)}. \end{aligned}$$

Note that like $L(W)$, $C(x)$ is a rational number in $[0,1]$. The more edges that are present in $N(x)$ (that is, the number of triangles in G incident with x), the larger is $C(x)$. The *clustering coefficient of* G, written $C(G)$, is the average of the clustering coefficients over all vertices, and so equals

$$\frac{1}{t}\sum_{x \in V(G)} C(x) = \frac{2}{t}\sum_{x \in V(G)} \frac{|E(G \restriction N(x))|}{\deg(x)(\deg(x)-1)}.$$

For example, a complete graph has clustering coefficient 1, while a triangle-free graph has clustering coefficient 0. See Figure 2.6 for another example. Adamic [**2**] reported that $C(W)$ (using a sample of W with around 250,000

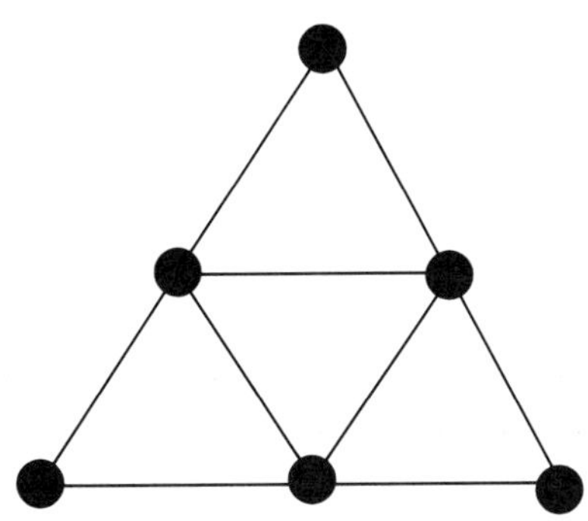

Figure 2.6. A graph H with $c(H) = 0.75$.

vertices) was 0.1078, in contrast to 2.3×10^{-4} for a randomly generated graph with the same number of vertices and edges.

2.1.4. Community structure. The web contains many *communities*: sets of pages sharing a common interest or topic. However, there is no consensus for a precise definition of a community in the web graph. Several approaches have been proposed, and we survey a few of them here.

A notion presented in Kleinberg et al. [**141**] and Kumar et al. [**146**] is that communities in the web are characterized by dense directed bipartite subgraphs. A *bipartite core* is a directed graph which contains at least one directed bipartite clique as a subgraph, where the directed edges in the subgraph all have tails of one fixed vertex class. See Figure 2.7 for an example. A motivation for this approach comes from viewing a community as being represented by pages with interest in a specific topic σ. The two vertex classes of the bipartite core represent good hub and authorities for σ. Hubs and authorities will be discussed in connection to the HITS web ranking algorithm in Chapter 5. In their study of communities in W, the

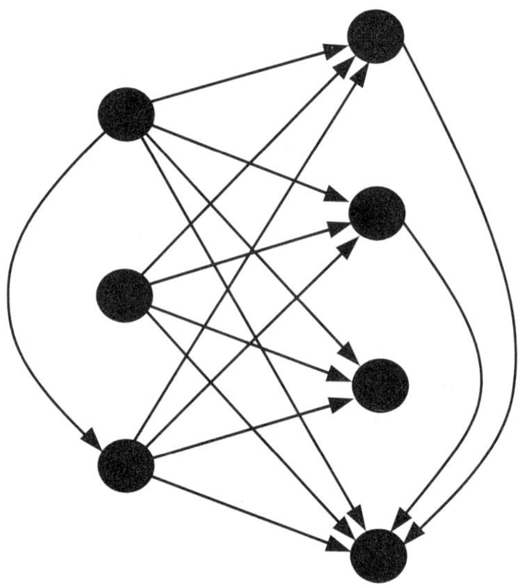

Figure 2.7. A bipartite core.

authors in [**141, 146**] show the presence of more small bipartite cores in W than a directed random graph with the same number of vertices and edges. A smaller bipartite core might indicate an emerging topic, while a larger one indicates a more established topic.

An alternative approach to communities was given by Flake, Lawrence, and Giles [**104**]. In their approach, a community is simply a subgraph which has more internal links than external ones. The authors use a max-flow min-cut formulation to efficiently identify communities. In a more recent work [**15**], an algorithm is presented for extracting communities as connected components of certain local graphs. The communities found by the algorithm have rich local structures (other than just bipartite cliques) and are highly locally connected.

2.2. Other Real-World Self-Organizing Networks

Around the same time in the late 1990's that research on properties of W began to intensify, there was a surge of research on experimental properties of other real-world graphs with similar properties. These networks arise naturally in many diverse disciplines (such as biology, computer science, and social science), but all share the common characteristics of being massive, sparse graphs with a power law degree distribution and small world structure. Such networks have now been coined *scale-free*, *complex*, *self-organizing*, or *heterogeneous*. Our own preferred term is *self-organizing*, since this is suggestive of the generative mechanisms underlying the network: over time, vertices choose their neighbour according to their own predisposition, rather than necessarily following global rules. Our focus in this book is W, but we would be remiss not to at least summarize some of the key facts of other self-organizing networks.

The networks we discuss fall into three main categories:

(1) technological;

(2) social; and

(3) biological.

As self-organizing networks are now under intense research scrutiny, we make no claims that our summary is exhaustive. The main message here is that the mathematical study of W to some extent overlaps with other research disciplines. A caveat is, despite the similarities, self-organizing networks usually have some properties distinct from W.

2.2.1. Technological networks. The most actively studied technological network is W, but it is by no means the only one. We describe some of the other self-organizing technological networks in this subsection. One of the earliest researched technological network is the *call graph*, whose vertices are telephone numbers, and two numbers are joined if one makes a long-distance call to the other over different intervals (such as over one day). The call graph is a directed graph. Aiello, Chung, and Lu [**5**] examined data from [**1**] that generated a 47 million vertex graph with around 8×10^7 edges. In [**5**], it was reported that the call graph was a power law graph with exponent $\beta = 2.1$.

The internet may be studied at various levels of structural organization. At the *interdomain level*, vertices are domains (which are sets of network addresses), and edges are interdomain connections. At the *router level*, vertices are routers (devices which transfer packets of data over the internet) and edges are interconnections between routers. These *internet networks* were first studied by Faloutsos, Faloutsos, and Faloutsos [**102**] who found power law degree distributions at the router level and interdomain level both with exponent $\beta = 2.48$. Although internet graphs are power law graphs, as reported in [**9**] they behave radically differently than W. For instance, the internet graphs are much less dynamic than W, and there are many more constraints owing to hardware and economic considerations. We will come back to a discussion of internet graphs in Section 4.3 of Chapter 4.

The *blog graph* or *Blogspace* is the digraph consisting of web blogs and the links between them. Blogspace is an induced subgraph of W. Graph-theoretical properties of Blogspace were first studied in [**147**], where power law in- and out-degree distributions with exponents close to 2 were discovered. The authors also noted that Blogspace contains a giant connected component and strong community structure.

2.2.2. Social networks. One of the first studies of empirical social networks was by Milgram [**162**], who estimated the diameter of the *friendship graph* (where vertices are people, and there is an edge between them if they are friends) to be 6. From this work, the phrase "Six degrees of separation" has become part of common parlance (for example, see Guare's play of the same name [**124**]).

Redner [**179**] studied the directed *network of citations* for all publications in Physical Review journals from the start of the journal in July 1893 until June 2003. Redner found a power law with exponent $\beta = 3$.

The *collaboration graph* consists of people in some discipline and undirected edges if they undertake some collaboration. For example, one collaboration graph consists of mathematicians with edges representing co-authorship. The famous *Erdős number* of a mathematician is the distance to the Paul Erdős vertex in the mathematics co-author graph. For example, my Erdős number is two: I co-authored a paper with P. Cameron, who co-authored with Erdős, while I did not co-author directly with Erdős. For more on the Erdős number, the reader is directed to [**123**].

Newman [**167**] studied the collaboration graph amongst biomedical researchers, physicists, and computer scientists. He reports that the average distance in these collaboration graphs is about 6, the diameter is about 20, and the clustering coefficient is at least 0.3. It is interesting to note that the degree distribution in this case does not follow a perfect power law, but rather a power law with exponential cut-off. Another collaboration graph is the *actor graph*, where vertices are actors (say in Hollywood), and two actors are joined if they appear in the same movie. The *Six Degrees of Kevin Bacon* game is played in the actor graph, with players trying to find paths (or shortest paths) to the Kevin Bacon vertex. Barabási and Albert [**21**] report a power law in the actor graph with exponent $\beta = 2.3$. Watts and Strogatz [**194**] found the average distance in the actor graph to be 3.65 and the clustering coefficient to be 0.79.

2.2.3. Biological networks. Unlike technological or social networks, biological networks do not necessarily model human activity, but often depict structure at the cellular level. A large number of biological networks have been recently studied, such as metabolic reaction networks, gene regulatory networks, and food networks between species in an ecosystem (see [**88**]). An important example of these networks is *protein-protein interaction* (*PPI*) networks, where vertices are proteins in a living cell, and undirected edges correspond to physical interactions between them. Proteins are central to cell function, and regulate and control several cellular functions by interacting with other proteins, enzymes and DNA. Interaction data for PPI are found by a number of methods, such as yeast 2-hybrid systems, and protein complex purification methods using mass spectrometry (see [**175**]).

The authors of [**87, 188**] studied the PPI in yeast and found a power law with exponent $\beta = 1.6$. Note that the exponent in the case of PPI's is less than 2, which is not the case for either technological or social networks. PPI networks tend to have giant connected components and the small world property. Biological evolution is the dominant decision making force in the generation of PPIs and other biological networks, so these graphs possess a radically different time-scale than W. For more information on PPIs, the reader is directed to [**69**] and [**175**].

Exercises

(1) Let X be a random variable with codomain $\mathbb{N} \setminus \{0\}$ satisfying a power law distribution with exponent $\beta > 1$. Show that X has finite expectation if and only if $\beta > 2$, and finite variance if and only if $\beta > 3$.

(2) Compute $C(G)$ and $L(G)$ if G is one of the following graphs, where n is a positive integer:
  (a) the Petersen graph;
  (b) a path P_n;
  (c) a clique K_n;
  (d) a cycle C_n;
  (e) a wheel W_n.

(3) (Requires knowledge of random graphs; see Chapter 3.) Compute asymptotic expressions for the expected values of $C(G)$ and $L(G)$ for $G \in G(n,p)$, where p is fixed in $(0,1)$.

(4) An alternative definition of the clustering coefficient due to [**35**] is the following:

$$C'(G) = \frac{\sum_{x \in V(G)} \binom{\deg(x)}{2} C(x)}{\sum_{x \in V(G)} \binom{\deg(x)}{2}}.$$

(a) Verify that $C'(G)$ is 3 times the ratio of the number of triangles in G to the number of pairs of adjacent edges.
(b) Compute $C'(G)$ for the graphs G in Exercise 2.

(5) Determine if the following (undirected) networks are power law graphs, and heuristically justify your answer.
(a) *The highway graph.* Vertices are cities, while edges correspond to roads between them.
(b) *The airport graph.* Vertices are airports, and edges are airline routes.
(c) *The musician graph.* Vertices are musicians, and edges correspond to having played together in a concert or recording.
(d) *The word graph.* Vertices are words in the English language, and two words are joined if they appear in the same sentence.
(e) *The board graph.* Vertices are people, with edges between those on the same corporate board.
(f) *The art gallery graph.* Vertices are art galleries, and two galleries are joined if they display art by the same artist.
(g) *The sex graph.* Vertices are people, and edges correspond to sexual relations.

(6) We explore a deterministic family of power law graphs defined in [**22**]. Consider the sequence $(G_t : t \in \mathbb{N})$ of graphs defined inductively as follows.

Let G_1 consist of K_1, which we call the *root*, and is labelled as a *special* vertex. Assume that G_t is defined, along with its set of special vertices. To form G_{t+1}, add two distinct isomorphic copies A and B of G_t to G_t. The special vertices of G_{t+1} are the union of the special vertices from A and B. Join each special vertex of A and B to the root vertex.
(a) Draw the graph G_3.
(b) Determine the number of vertices and edges of G_t for $t \geq 1$.
(c) Prove that the degree distribution of G_t follows a power law degree distribution with exponent $\beta = 1 + \frac{\log 3}{\log 2}$.

Chapter 3

Random Graphs

God does not play dice.

–Albert Einstein

So God does play dice with the universe. All the evidence points to him being an inveterate gambler, who throws the dice on every possible occasion.

–Stephen Hawking

3.1. Introduction

Graphs and probability mix well. Random graphs are important in the study of the web graph, and are a central tool in modern graph theory and theoretical computer science. The models of W that we will study in Chapter 4 incorporate randomness in their design. Knowledge of random graphs is, therefore, an important adjunct to understanding web graph models. One purpose of this chapter is to give the reader some background on the subject of random graphs that will be useful for our discussion of the models. The classical random graphs $G(n, p)$ do not satisfy properties observed in W; in particular, they are not power law graphs, but instead have binomial degree distributions. (See Theorem 3.11.) In any case, we will see that random graph theory is a beautiful subject in its own right.

Graph theory contains a myriad of elegant proofs using probabilistic methods, especially when other techniques are not applicable. An advantage of randomized methods is their ability to prove the existence of some object without explicitly constructing it. We will work only with discrete probability spaces, where, for example, integrals are replaced by sums.

Random graph theory began with the paper [**98**] of Paul Erdős and Alfréd Rényi. The subject of random graphs is now large and developed, spanning thousands of research papers. We will focus our efforts on exploring some fundamental methods useful for analyzing web graph models. These methods include the first and second moment method discussed in Sections 3.3 and 3.4, respectively, and martingales presented Section 3.5. For additional discussion of random graph theory see [**12**], [**32**], and [**137**].

3.2. What is a Random Graph?

We may define a probability space on graphs of a given order $n \geq 1$ as follows. Fix a vertex set V consisting of n distinct elements, usually taken as $[n] = \{1, 2, \ldots, n\}$, and fix $p \in [0, 1]$. Define the space of *random graphs of order n with edge probability p*, written $G(n, p)$, with sample space equalling the set of all $2^{\binom{n}{2}}$ (labelled) graphs with vertices V, and

$$\mathbb{P}(G) = p^{|E(G)|}(1-p)^{\binom{n}{2}-|E(G)|}.$$

Informally, we may view $G(n, p)$ as graphs with vertex set V, such that two distinct vertices are joined independently with probability p. Hence, V does not change, but the number of edges is not fixed: it varies according to a binomial distribution with expectation $\binom{n}{2}p$. Despite the fact that $G(n, p)$ is a space of graphs, we will abuse language and call it the *random graph of order n with edge probability p*. An alternative and equivalent definition of $G(n, p)$ using product spaces may be found in Exercise 21 of Chapter 1. See Figure 3.1.

An alternative to $G(n, p)$ is the space $G(n, M)$, which is the uniform probability space with sample space the set of all graphs on $[n]$ with exactly M edges. Hence, $\mathbb{P}(G)$ equals

$$\binom{\binom{n}{2}}{M}.$$

If M is close to $p\binom{n}{2}$, then in some cases the two spaces have similar properties (although the relation between the spaces is not trivial in general). See [**31**] for a discussion of $G(n, M)$ and other random graph processes.

We illustrate the power of random graphs via the following classical problem in pure graph theory. Given an integer $n \geq 1$, the *Ramsey number* $R(n)$ is the least integer m such that any graph of order m contains K_n or $\overline{K_n}$ as induced subgraphs. (This can be rephrased as follows: for each 2-colouring of the edges of a complete graph, there is a monochromatic clique of order n.) The numbers $R(n)$ are named after Frank Ramsey whose seminal 1930 paper [**178**] introduced similar combinatorial notions in the framework of mathematical logic. Ramsey theory, as it is now called, is

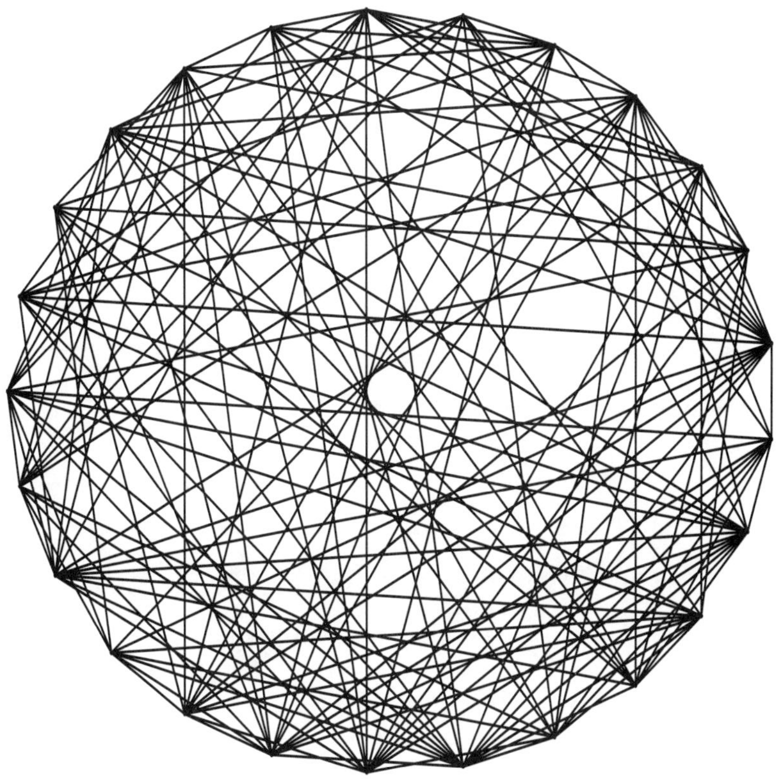

Figure 3.1. A graph with 25 vertices, where edges were drawn with probability $\frac{1}{2}$.

a broad and deep area of combinatorics. The numbers $R(n)$ exist by the following theorem (see Exercise 1(b)).

THEOREM 3.1. *For all* $n \geq 1$, $R(n) \leq 2^{2n-3}$.

By direct checking, $R(3) = 6$ and $R(4) = 18$. The exact values of $R(n)$, where $n \geq 5$, are not known. For example, it is known that $R(5)$ is in the interval $[43, 49]$ (see the dynamic survey [**177**] on small Ramsey numbers). An early success of random graphs came with a theorem of Erdős proved in 1947 [**94**] on lower bounds of Ramsey numbers.

THEOREM 3.2. *For each integer* $n \geq 3$, $R(n) > 2^{n/2}$.

Proof. We prove that for a fixed $m \leq 2^{n/2}$, with positive probability, $G \in G(m, \frac{1}{2})$ satisfies $\alpha(G) < n$ and $\omega(G) < n$. Hence, there is a graph of order m containing neither K_n nor its complement and the proof follows.

The probability that a given n-set S is a clique in $G \in G(m, \frac{1}{2})$ is $\left(\frac{1}{2}\right)^{\binom{n}{2}}$. As there are $\binom{m}{n}$ choices for S, we have that

$$\begin{aligned} \mathbb{P}(\omega(G) \geq n) &\leq \binom{m}{n}\left(\frac{1}{2}\right)^{\binom{n}{2}} \\ &\leq \frac{m^n}{2^n} 2^{-\frac{1}{2}(n^2-n)} \qquad (3.1) \\ &\leq 2^{\frac{n^2}{2}-n-\frac{1}{2}(n^2-n)} = 2^{-\frac{n}{2}} < \frac{1}{2}, \end{aligned}$$

as $n \geq 3$.

An analogous calculation shows that

$$\mathbb{P}(\alpha(G) \geq n) < \frac{1}{2}. \qquad (3.2)$$

Hence,

$$\mathbb{P}(\omega(G) < n \text{ and } \alpha(G) < n) = 1 - \mathbb{P}(\omega(G) \geq n \text{ or } \alpha(G) \geq n) > 0. \qquad \square$$

Erdős [**94**] in fact proved the stronger result that

$$R(n) > \frac{n}{e\sqrt{2}} 2^{n/2}$$

using Stirling's formula. As simple as the proof of Theorem 3.2 appears, no one has found a lower bound with better exponent. In particular, it is an open problem to find a real number $c > 0$ such that

$$R(n) > 2^{\left(\frac{1}{2}+c\right)n}.$$

The best known lower bound is

$$(1 + o(1))\frac{\sqrt{2}n}{e} 2^{n/2} < R(n)$$

proved in [**186**] using the Lovász local lemma.

The usual approach to $G(n,p)$ is *asymptotic*: we consider which properties hold for $G \in G(n,p)$ with high probability as n tends to infinity. To be precise, let $\mathcal{P}$ be a graph property (that is, a property preserved by isomorphism; for example, being planar and possessing Hamilton cycles are graph properties). We say that $G \in G(n,p)$ satisfies $\mathcal{P}$ *asymptotically almost surely* (or *a.a.s.* for short) if

$$\lim_{n\to\infty} \mathbb{P}(G \in G(n,p) \text{ satisfies } \mathcal{P}) = 1. \qquad (3.3)$$

If $\mathcal{P}$ satisfies condition (3.3), then we also say that $G \in G(n,p)$ satisfies $\mathcal{P}$ *with high probability (w.h.p)*. The edge probability p is often a function of n, written $p(n)$ (or simply p). The study of which properties are satisfied a.a.s. as a function of p is a central topic in random graph theory.

As an example of properties satisfied asymptotically by $G(n,p)$, we first consider adjacency properties. Adjacency properties of graphs were first discovered and investigated by Erdős and Rényi [**100**]. An *adjacency property* is a global property of a graph asserting that for every set S of vertices of some fixed type, there is a vertex joined to some of the vertices of S in a prescribed way. The so-called n-e.c. adjacency property has received much recent attention. For a positive integer n, a graph is *n-existentially closed* or *n-e.c.*, if for all disjoint sets of vertices A and B with $|A \cup B| = n$ (one of A or B can be empty), there is a vertex z not in $A \cup B$ joined to each vertex of A and no vertex of B. We say that z is *correctly joined* to A and B. Hence, for all n-subsets S of vertices, there exist 2^n vertices joined to S in all possible ways.

For example, a 1-e.c. graph is one with neither isolated nor universal vertices. A graph is 2-e.c. if for each pair of distinct vertices a and b, there are four vertices not equalling a and b joined to them in all possible ways. See Figure 3.3 for an example of a 2-e.c. graph. Although the n-e.c. property is straightforward to define, it is not obvious from the definition that graphs with the property exist. However, the existence of such graphs follows from our next result proved in [**100**].

THEOREM 3.3. *Fix an integer $n > 1$.*

(1) *A.a.s. $G \in G(m, \frac{1}{2})$ satisfies the n-e.c. property.*
(2) *Let f be a positive integer-valued function defined by*

$$f(m) = \binom{m}{n}\left(1 - \frac{1}{2^n}\right)^{m-n}.$$

If m is chosen so that $f(m) < 1$, then there is an n-e.c. graph of order m.

Proof. For item (1), fix disjoint sets of vertices A and B in G such that $A \cup B$ has cardinality n. For a given $z \notin (A \cup B)$, the probability that z is not correctly joined to A and B is

$$1 - \frac{1}{2^n}.$$

The probability that no vertex of G is correctly joined to A and B is therefore

$$\left(1 - \frac{1}{2^n}\right)^{m-n}.$$

As there are $\binom{m}{n}$ choices of A and B, the probability that G is not n-e.c. is at most $f(m)$, which tends to 0 as $m \to \infty$. For (2), by the choice of m and the last sentence, $G(m, \frac{1}{2})$ with positive probability contains an n-e.c. graph. $\square$

Theorem 3.3 implies that there are many examples of n-e.c. graphs (note also that it easily generalizes by replacing $\frac{1}{2}$ with any fixed real number $p \in (0,1)$; see Exercise 2). By Theorem 3.3 (2), if

$$m \geq 2^n n(n \log 2 + 2 \log n),$$

then there is an n-e.c. graph of order m. Despite these facts, until recently only one explicit family of n-e.c. graphs was known: the Paley graphs (which we will discuss below). This paradoxical quality of n-e.c. graphs being both common and rare has intrigued many researchers with different backgrounds such as graph theorists, logicians, design theorists, probabilists, and geometers.

For each positive integer n, define $m_{ec}(n)$ to be the minimum order of an n-e.c. graph. It is straightforward to see that $m_{ec}(1) = 4$. There are exactly three non-isomorphic 1-e.c. graphs of order 4: $2K_2, C_4$, and P_4; see Figure 3.2. If G is 2-e.c., then $G \restriction N(x)$ and $G \restriction N^c(x)$ are 1-e.c. Hence,

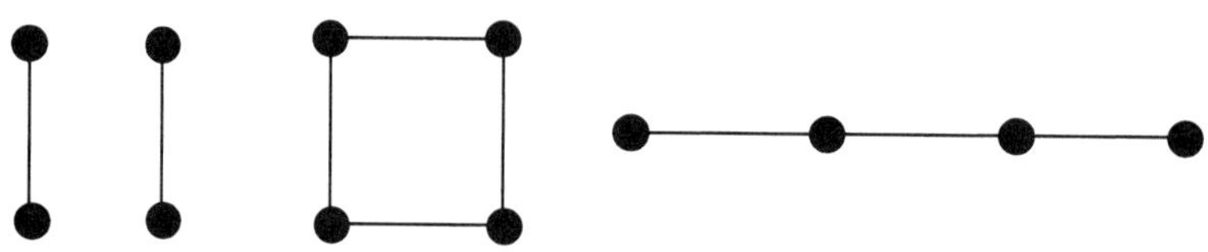

Figure 3.2. The 1-e.c. graphs of order 4.

$m_{ec}(2) \geq 9$. For two graphs G and H, the *Cartesian product* of G and H, written $G \square H$, has vertices $V(G) \times V(H)$ and edges $(a,b)(c,d) \in E(G \square H)$ if and only if $ac \in E(G)$ and $b = d$, or $a = c$ and $bd \in E(H)$. The notation comes from the fact that

$$K_2 \square K_2 \cong C_4.$$

The graph $K_3 \square K_3$ is shown in Figure 3.3. In [**57**] it was first noted that

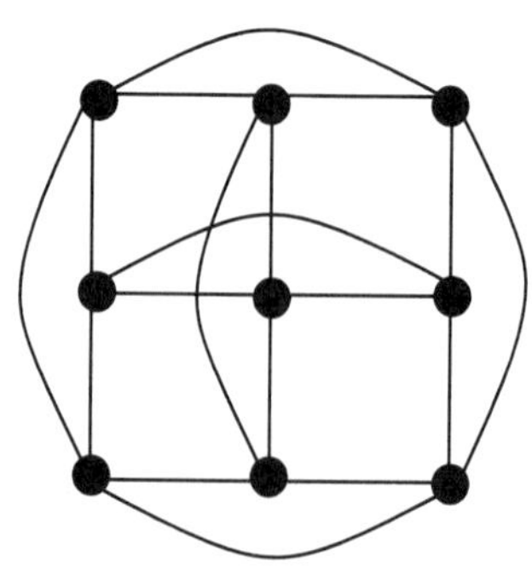

Figure 3.3. The graph $K_3 \square K_3$.

$m_{ec}(2) = 9$ and $K_3 \Box K_3$ is 2-e.c. In [**41**] it was observed that $K_3 \Box K_3$ is in fact the unique isomorphism type of 2-e.c. graphs of order 9 (see Exercise 2). In [**41**], it was noted that

$$20 \leq m_{ec}(3) \leq 28.$$

The exact determination of $m_{ec}(n)$, where $n \geq 3$, is a difficult open problem (how reminiscent of the situation for Ramsey numbers).

The n-e.c. condition may be viewed as a measure of randomness in deterministic graphs. This will be made clearer in Chapter 6 when we discuss the infinite random graph (which is the unique isomorphism type of countable graphs that is n-e.c. for all positive integers n). Other notions of random-like behaviour in deterministic graphs were proposed and thoroughly investigated. Two such notions are *quasi-randomness* [**70**] and *pseudo-randomness* [**190**]. However, these randomness properties do not necessarily imply the n-e.c. properties. Examples are given in [**63**] which are pseudo-random but not 4-e.c. For additional information on pseudo-randomness, the interested reader is directed to the surveys [**68**] and [**143**].

We describe in detail the quasi-randomness properties first discovered for graphs by Chung, Graham, and Wilson [**70**]. These properties all hold a.a.s. in $G(n, \frac{1}{2})$ (see Exercise 4). A remarkable fact is that a deterministic family of graphs satisfying one of the properties satisfies all of them.

To state the properties, we require some notation. As we will further discuss in Chapter 5, to each graph G of order n we may associate its $n \times n$ adjacency matrix $A(G)$. As $A(G)$ is a real-symmetric matrix, it has n real eigenvalues $\lambda_1 > 0, \lambda_2, \ldots, \lambda_n$, which are ordered by absolute value:

$$\lambda_1 = |\lambda_1| \geq \cdots \geq |\lambda_n|.$$

Let $N_S(H, G)$ be the number of labelled subgraphs of G isomorphic to H, while $N_{IS}(H, G)$ is the number of labelled induced subgraphs of G isomorphic to H. For $X \subseteq V(G)$, let $e(X) = |E(G \restriction X)|$. For vertices u and v, define $s(u, v)$ to be the set of vertices joined to both u and v or neither u nor v. For a fixed infinite family $\{G_n : n \in \mathbb{N}\}$ where G_n has order n, we define the following five properties.

(P1) For all $X \subseteq V(G_n)$

$$e(X) = \frac{1}{4}|X|^2 + o(n^2).$$

(P2) $e(G) \geq (1 + o(1))\frac{n^2}{4}$, and $N_S(C_4, G_n) \leq (1 + o(1))\frac{n^4}{16}$.

(P3) $e(G) \geq (1+o(1))\frac{n^2}{4}$, and for any fixed graph H of order $4 \leq m \leq n$,

$$N_{IS}(H, G_n) = (1 + o(1))n^t 2^{-\binom{t}{2}}.$$

(P4) $\sum_{u,v \in V(G_n)} \left| |N(u) \cap N(v)| - \frac{n}{4} \right| = o(n^3)$.

(P5) $\sum_{u,v \in V(G_n)} \left| s(u,v) - \frac{n}{2} \right| = o(n^3)$.

(P6) $e(G) \geq (1 + o(1))\frac{n^2}{4}$ and for $2 \leq i \leq n$

$$\lambda_1 = (1 + o(1))\frac{n}{2}, \; \lambda_n = o(n).$$

The following theorem was proved in [**70**].

THEOREM 3.4 ([**70**]). *If G_n satisfies any one of the six properties above, then it satisfies all of them.*

Analogues of Theorem 3.4 have been considered for a wide variety of combinatorial structures, such as sparse graphs [**73**], tournaments [**72**], hypergraphs [**71, 67**], permutations [**83**], and most recently, for power law graphs [**74**]. We will consider quasi-random power graphs in Chapter 4.

Most of the known explicit quasi-random and n-e.c. graphs are strongly regular. A k-regular graph G with v vertices, such that each pair of joined vertices has exactly λ common neighbours, and each pair of non-joined vertices has exactly μ common neighbours is called a *strongly regular graph*; we say that G is an $\mathrm{SRG}(v, k, \lambda, \mu)$. The first family of explicit graphs that was discovered to contain n-e.c. members for all n was Paley graphs. Paley graphs are defined over certain finite fields, and it has long been observed that they satisfy many of the properties of the random graph $G(n, \frac{1}{2})$. The *Paley graph* of order q, for a prime power $q \equiv 1 \pmod 4$, is the graph P_q whose vertices are the elements of the finite field $\mathrm{GF}(q)$ in which two distinct vertices x and y are joined if and only if $x - y$ is a square in $\mathrm{GF}(q)$. For example, P_5 is isomorphic to the 5-cycle, while P_9 is isomorphic to the graph $K_3 \Box K_3$ of Figure 3.3.

Paley graphs have many properties as described in the following theorem, which may be proved using properties of finite fields (for a proof see, for example, Chapter 13 of [**32**]).

THEOREM 3.5. *Fix q a prime power with $q \equiv 1 \pmod 4$.*

(1) The graph P_q is an $\mathrm{SRG}(q, \frac{q-1}{2}, \frac{q-5}{4}, \frac{q-1}{4})$.

(2) The graph P_q is *self-complementary*; that is, $P_q \cong \overline{P_q}$.

(3) The graph P_q is *symmetric*; that is, it is vertex- and edge-transitive.

The Paley graphs are quasi-random, as first noted in [**70**].

THEOREM 3.6. *The Paley graphs satisfy property (P5) and so are quasi-random.*

Proof. Fix vertices u and v in P_q. Then z is joined to u and v or joined to neither u nor v if and only if

$$\frac{z-u}{z-v}$$

is a square (mod q). There are $\frac{1}{2}(q-1)-1=\frac{1}{2}(q-3)$ many squares in $\mathrm{GF}(q)$ distinct from 1. Let a be such a square. As there is a unique solution to

$$\frac{z-u}{z-v}=1+\frac{v-u}{z-v}=a,$$

it follows that $s(u,v)=\frac{1}{2}(q-3)$. Hence,

$$\begin{aligned}\sum_{u,v\in V(G_n)}\left|s(u,v)-\frac{q}{2}\right| &= \sum_{u,v\in V(G_n)}\left|\frac{1}{2}(q-3)-\frac{q}{2}\right| \\ &= \binom{q}{2}\frac{3}{2}=\frac{3\left(q^2-q\right)}{4} \\ &= o(q^3). \qquad \square\end{aligned}$$

The next result demonstrates that sufficiently large Paley graphs are n-e.c.

THEOREM 3.7 ([**29, 38**]). *If $q>n^2 2^{2n-2}$, then P_q is n-e.c.*

The proofs of Theorem 3.7 in [**29, 38**] each use a famous result from number theory on character sum estimates, namely Weil's proof of the Riemann hypothesis over finite fields. For a proof of the following theorem and additional background, see [**180**].

THEOREM 3.8 ([**180**]). *Let χ be a non-trivial character of order d over* $\mathrm{GF}(q)$. *Suppose that $f(x)$ is a polynomial over* $\mathrm{GF}(q)$ *with exactly m distinct zeros and is not of the form $c(g(x))^d$, where $c\in$* $\mathrm{GF}(q)$ *and $g(x)$ is a polynomial over* $\mathrm{GF}(q)$. *Then*

$$\left|\sum_{x\in\mathrm{GF}(q)}\chi(f(x))\right|\leq(m-1)q^{1/2}.$$

We now give a new randomized construction of explicit n-e.c. graphs. This method generates strongly regular n-e.c. graphs, and requires a short, but pleasant detour through finite geometry. Consider an *affine plane A of order q*, where A is coordinatized over $\mathrm{GF}(q)$, with q a prime power. In particular, A is a 2-$(q^2,q,1)$ design (with blocks called *lines*), and hence, satisfies the property that given a point x and a line ℓ, there is a unique line parallel to ℓ that goes through x. Each line contains exactly q points. The relation of parallelism on the set of lines is an equivalence relation, and the equivalence classes are called *parallel classes.* There are $q+1$ parallel classes (corresponding to points on the line at infinity). We use the notation pq for the line between p and q. Although this notation conflicts with our earlier notation for edges of a graph, we keep both notations since they are standard.

We now consider a construction of strongly regular graphs which is due to Delsarte and Goethals, and to Turyn; see [**181**]. Let ℓ_∞ be the $(q+1)$-element line at infinity, identified with slopes. Fix $S \subseteq \ell_\infty$. Define $G(q,S,A)$ to have vertices the points of A, and two vertices p and q are joined if and only if the line pq has slope in S. The graph $G(q,S,A)$ is a

$$\mathrm{SRG}(q^2, |S|(q-1), q-2+(|S|-1)(|S|-2), |S|(|S|-1))$$

(see Exercise 5). Let $\mathcal{G}(q,A)$ be the family of graphs $G(q,S,A)$ for all choices of S; if $0 \le k \le q+1$ is fixed, then write $\mathcal{G}(q,k,A)$ for the subfamily of all graphs in $\mathcal{G}(q,A)$ where $|S|=k$. For a fixed k, $\mathcal{G}(q,k,A)$ may contain non-isomorphic members (in general, this is non-trivial to determine; see [**61**]).

Let q be a power of a fixed prime, and let A be an affine plane with q points coordinatized over the field with q elements, written $\mathrm{GF}(q)$. For a fixed $p \in (0,1)$, choose $m \in \ell_\infty$ to be in S independently with probability p; with the remaining probability, m is in the complement of S. This makes $\mathcal{G}(q,A)$ into a probability space which we denote $\mathcal{G}_p(q,A)$. While $|S|$ is a random variable, all choices of S give rise to a strongly regular graph. We prove that for large q, the space $\mathcal{G}_p(q,A)$ contains n-e.c. graphs.

THEOREM 3.9. *Fix $p \in (0,1)$, and let n be a positive integer. Then a.a.s. $G \in \mathcal{G}_p(q,A)$ is n-e.c.*

Proof. Fix $G \in \mathcal{G}_p(q,A)$, and fix disjoint sets of vertices X and Y in G, with $|X \cup Y| = n$. Let $U = X \cup Y$. We prove that for large q, with probability 1 there is a vertex z correctly joined to X and Y. To accomplish this, we construct a set P_U of points, disjoint from U, such that with probability 1, z is in P_U. We set $s = \lceil q^b \rceil$, where b is chosen so that $0 < b < \frac{1}{2}$.

Fix a point v of A. The *projection from v onto l_∞* is the map

$$\pi_v : A \backslash \{v\} \to \ell_\infty$$

taking a point x to the intersection of vx with ℓ_∞. Hence, $\pi_v(x)$ is the slope of the line vx. If V is a set of points, then let $\pi_v(V) = \bigcup_{x \in V} \pi_v(x)$.

For sufficiently large q, we inductively construct a set of points P_U distinct from U with the following properties.

(1) If $p \in P_U$, then $|\pi_p(U)| = n$.

(2) For all distinct p and q in P_U, $\pi_p(U) \cap \pi_q(U) = \emptyset$.

(3) $|P_U| = s$.

Define $P_{U,1}$ by choosing any point $p_1 \notin U$ that is not on a line joining two points of U. As

$$n + \binom{n}{2}(q-2) = o(q^2),$$

we may find such a p_1 for large q.

For a fixed $i \geq 1$, suppose that $P_{U,i}$ has been constructed for large q, with $P_{U,i}$ containing $P_{U,1}$, and $|P_{U,i}| = i$. We would like to choose $p_{i+1} \notin U$ to be a point that is

(i) not on a line joining two points of U, and

(ii) not on a line joining a point of U to a point in $\bigcup_{j=1}^{i} \pi_{p_j}(U)$.

Condition (i) rules out $\binom{n}{2}$ lines, while (ii) rules out $ni + n(n-1)i$ lines. As

$$n + \binom{n}{2}(q-2) + ni(q-1) + n(n-1)i(q-2) = o(q^2)$$

for $i \leq s$ by the choice of b, we may find a suitable p_{i+1} satisfying items (1) and (2) for large q. Add p_{i+1} to $P_{U,i}$ to form $P_{U,i+1}$. Define

$$P_U = \bigcup_{i=1}^{s} P_{U,i}$$

so $|P_U| = s$.

For a fixed $U = X \cup Y$, we estimate the probability that none of the vertices of P_U are correctly joined to X and Y. Suppose that $m = |X|$ and $n - m = |Y|$. By item (1), note that any z in P_U has the property that zx and zy have distinct slopes, where x, y are distinct points of U. Note also that zx is an edge of G if and only if $\pi_z(x) \in S$. Therefore, the probability that a given z in P_U is not joined correctly to X and Y is

$$1 - p^m(1-p)^{n-m}. \tag{3.4}$$

By item (2) in the defining properties of P_U, any two distinct points of P_U induce disjoint slope sets in ℓ_∞. In particular, the probability (3.4) independently holds for any choice of z in P_u. Hence, the probability that no z in P_U is correctly joined to X and Y is

$$(1 - p^m(1-p)^{n-m})^{\lceil q^b \rceil}.$$

Let α be the maximum of $(1 - p^m(1-p)^{n-m})$ over all choices of m. The probability that G is not n-e.c. is therefore at most

$$\binom{q^2}{n}\alpha^{\lceil q^b \rceil} = O(q^{2n}\alpha^{q^b}) = o(1),$$

as α is a fixed real number in $(0,1)$ and $0 < b$. $\square$

If q is odd and $|S| = \frac{q+1}{2}$, then it is an exercise to show that the graphs in $G(q, S, A)$ are quasi-random.

3.3. Expectation and the First Moment Method

A graph parameter X becomes a random variable on $G(n,p)$. For example, the size (that is, number of edges), chromatic number, and clique number are all random variables on random graphs. Computing the exact values of these parameters is often difficult or NP-complete for deterministic graphs, but computing their averages is sometimes easier. The use of expectation (sometimes called the *first moment method*) is an important tool in understanding properties of random variables in $G(n,p)$.

Linearity of expectation, although a basic fact in probability texts, has interesting consequences for random graphs. As an example, let X be the number of edges in $G(n,p)$. For each unordered pair $u = \{i,j\}$ from $[n]$, let X_u be the *indicator random variable* for u being an edge. That is,

$$X_u = \begin{cases} 1 & \text{if } ij \in E, \\ 0 & \text{else.} \end{cases}$$

Then $\mathbb{E}(X_u) = p$. As

$$X = \sum_u X_u,$$

by the linearity of expectation we have that

$$\mathbb{E}(X) = \binom{n}{2} p.$$

Markov's inequality (see Chapter 2) may be used to prove a multitude of results on $G(n,p)$. A startling application of this inequality is Erdős' theorem proved in [**95**] on graphs with large girth and chromatic number. The *girth* of a graph is the length of its smallest cycle. At first sight, we may think that a graph without small cycles would have low chromatic numbers. However, this is far from the case.

THEOREM 3.10. *For every integer $m \geq 2$, there is a graph G such that $\chi(G)$, $g(G) \geq m$.*

Proof. Consider $G \in G(n,p)$ for n a positive integer and $p = p(n)$ to be determined. As

$$\chi(G) \geq \frac{|V(G)|}{\alpha(G)},$$

our approach to showing that $\chi(G)$ is large is to show that $\alpha(G)$ is small when compared to n. For $2 \le i \le n$,

$$\begin{aligned}\mathbb{P}(\alpha(G) \ge i) &\le \binom{n}{i}(1-p)^{\binom{i}{2}} \\ &\le \left(n(1-p)^{\frac{i-1}{2}}\right)^i \qquad (3.5) \\ &\le \left(n\exp\left(-p\frac{i-1}{2}\right)\right)^i.\end{aligned}$$

Now choose $p = n^{-\frac{m}{m+1}}$. Choose n large enough so that $n^{\frac{1}{m+1}} \ge 6m\log n$. Hence, with $i = \lceil \frac{n}{2m} \rceil$ we have that

$$\begin{aligned}\mathbb{P}(\alpha(G) \ge i) &\le \left(n\exp\left(\frac{-pi}{2}\right)e^{p/2}\right)^i \\ &\le \left(n\exp\left(-\frac{3}{2}\log n\right)e^{1/2}\right)^i \\ &= \left(\left(\frac{e}{n}\right)^{\frac{1}{2}}\right)^i = o(1),\end{aligned}$$

where the first inequality follows from (3.5), and the second from the fact that $pi \ge 3\log n$.

Choose an integer n_1 so that for $n \ge n_1$,

$$\mathbb{P}\left(\alpha(G) \ge \frac{n}{2m}\right) < \frac{1}{2}. \qquad (3.6)$$

We now consider the girth of G. Let X be the random variable counting the number of cycles of length at most m in $G(n,p)$. Then (see Exercise 10)

$$\begin{aligned}\mathbb{E}(X) &= \sum_{i=3}^{m}\binom{n}{i}\frac{(i-1)!}{2}p^i \\ &\le \frac{1}{2}\sum_{i=3}^{m} n^i p^i \\ &\le \frac{1}{2}(m-2)(np)^m,\end{aligned}$$

since $np \ge 1$. By Markov's inequality, we have that

$$\begin{aligned}\mathbb{P}\left(X \ge \frac{n}{2}\right) &\le \frac{\mathbb{E}(X)}{n/2} \\ &\le (m-2)\frac{(np)^m}{n} \\ &= (m-2)n^{-\frac{1}{m+1}} = o(1),\end{aligned}$$

where the second inequality follows from the choice of p.

Choose an integer n_2 so that for $n \geq n_2$,

$$\mathbb{P}\left(X \geq \frac{n}{2}\right) < \frac{1}{2}. \tag{3.7}$$

Now for $n \geq n_1 + n_2$, we have by (3.6) and (3.7) that

$$\mathbb{P}\left(\alpha(G) \geq \frac{n}{2m} \text{ or } X \geq \frac{n}{2}\right) < 1. \tag{3.8}$$

Therefore, by (3.8) there exists a graph J on n vertices with $\alpha(J) < \frac{n}{2m}$ and $X(J) < \frac{n}{2}$. In J, delete one vertex from each cycle of length at most m to obtain the graph G of order n. Then $g(G) \geq m$, $|V(G)| \geq n/2$, and $\alpha(G) < \frac{n}{2m}$. Hence,

$$\chi(G) \geq \frac{\frac{n}{2}}{\frac{n}{2m}} = m. \qquad \square$$

Explicit examples of graphs with large girth and chromatic numbers were found by Lovász [**155**]. We explore a family of triangle-free graphs with large chromatic numbers in Exercise 11.

We now turn to vertex degrees in $G(n,p)$. We will use the stronger notion of w.e.p. in favour of the more commonly used a.a.s., since it simplifies some proofs. An event holds *with extreme probability* (*w.e.p.*) if it holds with probability at least $1 - \exp(-\Theta(\log^2 n))$ as $n \to \infty$. Observe that if we consider a polynomial number of events that each holds w.e.p., then w.e.p. all events hold. The following result is our first exposure to a concentration result in random graphs, stating that the degree of a vertex is asymptotically close to its expected degree.

THEOREM 3.11. *If $p \in (0,1)$ is fixed, then w.e.p. every vertex of $G \in G(n,p)$ has degree equal to*

$$pn + O(\sqrt{pn}\log n) = (1 + o(1))pn.$$

Proof. Let Y be the number of vertices with degree greater than $pn + \sqrt{pn}\log n$ or of degree smaller than $pn - \sqrt{np}\log n$. To prove the theorem, it is enough to show that the expected number $\mathbb{E}(Y)$ is tending to zero faster than the function $\exp(-c(\log^2 n))$ as $n \to \infty$, for some constant $c > 0$. The conclusion then follows from Markov's inequality

$$\mathbb{P}(Y \geq t) \leq \frac{\mathbb{E}(Y)}{t}.$$

Indeed,

$$\mathbb{P}(Y = 0) = 1 - \mathbb{P}(Y \geq 1) \geq 1 - \mathbb{E}(Y) > 1 - \exp(-\Theta(\log^2 n)).$$

Now consider a binomially distributed random variable $X \in Bi(n,p)$ with $\mathbb{E}(X) = np$. The following bounds, often called the *Chernoff bounds*, then hold (see Theorem 1.12 of Chapter 1 and also Theorem 2.1 in [**137**]):

$$\mathbb{P}(X \geq \mathbb{E}(X) + t) \leq \exp(-\mathbb{E}(X)\,\phi(t/\mathbb{E}(X))), \tag{3.9}$$

$$\mathbb{P}(X \leq \mathbb{E}(X) - t) \leq \exp(-\mathbb{E}(X)\,\phi(-t/\mathbb{E}(X))), \tag{3.10}$$

where $\phi(x) = (1+x)\log(1+x) - x$ for $x \geq -1$, and $\phi(x) = \infty$ for $x < -1$. We use the following two-sided estimate which is an immediate corollary of (3.9) and (3.10):

$$\mathbb{P}(|X - \mathbb{E}(X)| \geq \varepsilon\mathbb{E}(X)) \leq 2\exp\left(-\frac{1}{3}\varepsilon^2\mathbb{E}(X)\right), \tag{3.11}$$

where $\varepsilon \leq 3/2$.

Fix any vertex v of G. Then $\mathbb{E}(\deg(v)) = pn - O(1)$. Using (3.11) with $\varepsilon = 0.5\log n/\sqrt{\mathbb{E}(\deg(v))} = o(1)$, we have that

$$\mathbb{P}(|\deg(v) - \mathbb{E}(\deg(v))| \geq \varepsilon\mathbb{E}(\deg(v))) \leq \exp\left(-\Omega(\log^2 n)\right),$$

which goes to zero faster than any fixed polynomial.

It follows that

$$\mathbb{E}(Y) \leq n\exp\left(-\Omega(\log^2 n)\right) = \exp\left(-\Omega(\log^2 n)\right).$$

Thus, w.e.p. all vertices have degree $pn + O(\sqrt{pn}\log n)$. □

Theorem 3.11 can be extended to the case with p a function of n, where the situation becomes more complicated. If $\lim_{n\to\infty} np > 0$, then the degree distribution of $G(n,p)$ is asymptotically Poisson. For more details, see Chapter 3 of [**32**].

3.4. Variance and the Second Moment Method

The distribution of a random variable may vary a great deal from its expected value. The use of variance (or the *second moment method*) allows us to measure how much a random variable deviates from its expected value. Besides the expectation of a random variable X, the variance of X, written $\mathrm{Var}(X)$, is of great importance in random graph theory. For example, by Chebyshev's inequality, we have the following.

THEOREM 3.12. *Let X be a non-negative integer-valued random variable on $G(n,p)$. If*

$$\mathrm{Var}(X) = o(\mathbb{E}(X)^2),$$

then a.a.s. $X > 0$ and $X \sim \mathbb{E}(X)$.

Proof. By Chebyshev's inequality we have that

$$\mathbb{P}(X=0) \leq \frac{\operatorname{Var}(X)}{\mathbb{E}(X)^2},$$

which proves that a.a.s. $X > 0$. For every $\varepsilon > 0$, Chebyshev's inequality implies that

$$\mathbb{P}(|X - \mathbb{E}(X)| \geq \varepsilon \mathbb{E}(X)) \leq \frac{\operatorname{Var}(X)}{\varepsilon^2 \mathbb{E}(X)^2}$$

and so a.a.s. $X \sim \mathbb{E}(X)$. □

Theorem 3.12 is an example of a *concentration result*: the random variable X and $\mathbb{E}(X)$ are close as n becomes large. As an illustrative example of the second moment method, let us consider the number of triangles in $G(n,p)$ for various values of p.

THEOREM 3.13. (1) *If* $np = o(1)$, *then*

$$\lim_{n\to\infty} \mathbb{P}(G \in G(n,p) \text{ contains a } K_3) = 0.$$

(2) *If* $\frac{1}{np} = o(1)$, *then*

$$\lim_{n\to\infty} \mathbb{P}(G \in G(n,p) \text{ contains a } K_3) = 1.$$

Proof. Let X be the random variable on $G(n,p)$ equalling the number of distinct triangles. Then

$$\mathbb{E}(X) = \binom{n}{3} p^3 = O\left(n^3 p^3\right). \tag{3.12}$$

For item (1), note that if $np = o(1)$, then by (3.12) we have that $\mathbb{E}(X) = o(1)$. The assertion of (1) then quickly follows from Markov's inequality

$$\mathbb{P}(X \geq 1) \leq \mathbb{E}(X).$$

For (2), we compute $\operatorname{Var}(X)$. The work comes from computing $\mathbb{E}(X^2)$. By using indicator variables, it follows that

$$\mathbb{E}(X^2) = \sum_{(A,B)} \mathbb{P}(A, B \text{ are triangles}). \tag{3.13}$$

There are three cases of terms in (3.13).

i) A and B have 3 elements in common.

This case occurs with probability p^3 and occurs $\binom{n}{3}$ times in the sum (3.13).

ii) A and B have 2 elements in common.

This case occurs with probability p^5 and occurs $6\binom{n}{4}$ times in the sum (3.13).

iii) A and B have at most 1 element in common.

In this case, this occurs with probability p^6 and occurs $\binom{n}{3}^2 - \binom{n}{3} - 6\binom{n}{4}$ times in (3.13). Hence,

$$\begin{aligned}\mathbb{E}(X^2) &= \binom{n}{3}p^3 + 6\binom{n}{4}p^5 + p^6\left(\binom{n}{3}^2 - \binom{n}{3} - 6\binom{n}{4}\right) \\ &= \mathbb{E}(X)^2 + O(n^3p^3(1-p^3)) + O(n^4p^5(1-p)).\end{aligned} \tag{3.14}$$

By (3.14), we have that

$$\mathrm{Var}(X) = O(n^3p^3(1-p^3)) + O(n^4p^5(1-p)) = o(\mathbb{E}(X)^2).$$

By Theorem 3.12, a.a.s. $X > 0$. □

Theorem 3.13 is one fragment of a much larger class of results. Given a graph property $\mathcal{P}$ (which can be formally thought of as a class of graphs closed under isomorphism) we say that a non-zero real function $t(n)$ is a *threshold function for* $\mathcal{P}$ if

$$\lim_{n\to\infty} \mathbb{P}(G \in G(n,p) \text{ satisfies } \mathcal{P}) = \begin{cases} 1 & \text{if } t/p = o(1), \\ 0 & \text{if } p/t = o(1). \end{cases}$$

The following result was proved by Erdős and Rényi [**99**]. A graph G is *balanced* if for all subgraphs G' of G,

$$\frac{|E(G')|}{|V(G')|} \le \frac{|E(G)|}{|V(G)|}.$$

For example, a tree, cycle, or clique are balanced (Exercise 15).

THEOREM 3.14 ([**99**]). *If G is a non-trivial balanced graph with v vertices and e edges, then $t(n) = n^{\frac{-v}{e}}$ is a threshold function for the property of containing an isomorphic copy of G.*

More generally, Bollobás and Thomason [**51**] proved that all *increasing properties* (properties invariant under adding an edge to the graph) have threshold functions.

Using the first and second moment and other methods, the evolution (as $p = p(n)$ increases in order from 0 to 1) of $G(n,p)$ was studied by Erdős and Rényi [**98**]. We may view $G(n,p)$ as a growing organism, evolving from an empty to complete graph as p increases from 0 to 1. Detailed information and recent results about the evolution can be found in [**12**] and [**32**]. For instance, at $p = \frac{1}{n}$ a macroscopic change takes place in $G(n,p)$. When $p = \frac{c}{n}$ with $c < 1$, a.a.s. $G \in G(n,p)$ consists of small components, the largest of cardinality $\Theta(\log n)$. However, when $p = \frac{c}{n}$ with $c > 1$, a giant component of cardinality $\Theta(n)$ emerges, absorbing many of the smaller components. This remarkable phenomenon was called the *double jump* by Erdős and Rényi.

In the special case when the variance of a sequence of random variables converges to 0, we can derive a convergence (and therefore concentration) result for the sequence. A sequence of random variables $(X_n : n \in \mathbb{N})$ *converges in probability to the random variable* X if for all real numbers $\varepsilon > 0$,

$$\lim_{n\to\infty} \mathbb{P}(|X_n - X| \geq \varepsilon) = 0.$$

COROLLARY 3.15. *Let $(X_n : n \in \mathbb{N})$ be a sequence of random variables and X a random variable that does not depend on n so that $\lim_{n\to\infty} \mathbb{E}(X_n) = \mathbb{E}(X)$. If $\lim_{n\to\infty} \mathrm{Var}(X_n) = 0$, then $(X_n : n \in \mathbb{N})$ converges in probability to X.*

Proof. We have that

$$\begin{aligned}
\lim_{n\to\infty} \mathbb{E}(X_n - X)^2 &= \lim_{n\to\infty} \mathbb{E}(X_n^2) - 2\lim_{n\to\infty} \mathbb{E}(X_n)\mathbb{E}(X) + \lim_{n\to\infty} \mathbb{E}(X)^2 \\
&= \lim_{n\to\infty} \mathbb{E}(X_n^2) - \mathbb{E}(X)^2 \\
&= \lim_{n\to\infty} \mathbb{E}(X_n^2) - \mathbb{E}(X_n)^2 \\
&= \lim_{n\to\infty} \mathrm{Var}(X_n) = 0,
\end{aligned}$$

where the first equality follows by the linearity of expectation and since X is constant, and the second and third equalities follow from the hypothesis. Hence, $(X_n : n \in \mathbb{N})$ converges to X in the L^2 norm, which implies that $(X_n : n \in \mathbb{N})$ converges to X in probability. (See Theorem 3 in 7.2 of [**121**].) □

Corollary 3.15 will be useful in our study of the degree distribution of graphs generated by web graph models in Chapter 4.

3.5. Martingales and Concentration

Martingales are useful in random graph theory for proving concentration around the expected value. The Chernoff bounds apply to sums of independent random variables; martingale concentration is useful when there is dependence, or concentration using the variance is not otherwise applicable. To describe martingale concentration, we need some background material on conditional expectation. A rigorous discussion of conditional expectation will take us too deeply into measure theory. Our somewhat informal discussion below is useful for our purposes.

Let X, Y be r.v.'s on a common probability space such as $G(n,p)$. The *conditional mass function* of X given $Y = y$, written $f_{X|Y}(\cdot|y)$, is defined as

$$f_{X|Y}(x|y) = \mathbb{P}(X = x | Y = y),$$

for all y such that $\mathbb{P}(Y = y) > 0$. Given $Y = y$, we may think of $f_{X|Y}(x|y)$ as a function of x. The expected value of this distribution, which is

$$\sum_x x f_{X|Y}(x|y),$$

is the *conditional expectation of* X *when* $Y = y$, and is written

$$\mathbb{E}[X|Y = y].$$

Define $g(y) = \mathbb{E}[X|Y = y]$. The function g is the *conditional expectation of* X *on* Y, written $\mathbb{E}[X|Y]$. Note that $\mathbb{E}[X|Y]$ is a random variable, and so has an expected value. Intuitively, $\mathbb{E}[X|Y]$ is the expected value of X assuming Y is known. From the definitions (see Exercise 20) it follows that

$$\mathbb{E}(\mathbb{E}[X|Y]) = \mathbb{E}(X). \tag{3.15}$$

We may generalize this to $\mathbb{E}[X|Y_1, \ldots, Y_s]$, where the Y_i are random variables. The following property is useful (see Exercise 20).

THEOREM 3.16 (Tower property).

$$\mathbb{E}[\mathbb{E}[X|Y_1, \ldots, Y_s]|Z_1, \ldots, Z_t] = \mathbb{E}[X|Z_1, \ldots, Z_t].$$

A sequence $(X_i : 0 \leq i \leq t)$ of random variables over a common probability space is a *martingale* if for all $i \geq 1$,

$$\mathbb{E}[X_i|X_0, \ldots, X_{i-1}] = X_{i-1}.$$

A classic example of a martingale comes from a gambler playing a sequence of *fair games* (that is, the expected win or loss of each play is 0). Define X_i to be the gambler's capital after the ith game. Since the game is fair, the sequence $(X_i : 0 \leq i \leq t)$ is a martingale.

Martingales can in fact be derived from any sequence of random variables. To be more precise, let A and $(Z_i : 1 \leq i \leq t)$ be random variables on a common probability space; define $X_0 = \mathbb{E}(A)$ and let $X_i = \mathbb{E}[A|Z_1, \ldots, Z_i]$, where $1 \leq i \leq t$.

LEMMA 3.17. *The sequence of random variables $(X_i : 0 \leq i \leq t)$ as defined above is a martingale.*

We leave the proof as an exercise. The martingale $(X_i : 0 \leq i \leq t)$ is called the *Doob martingale of* A *with respect to the* Z_i. We can think of A as a function

$$f(Z_1, \ldots, Z_t)$$

determined by the random variables $(Z_i : 1 \leq i \leq t)$. Then $X_0 = \mathbb{E}(A)$ and $X_t = A$. The Doob martingale reveals more and more information about f over time, so in the end all information is available and we obtain the actual value of f.

A useful application of the Doob martingale comes with the edge- and vertex-exposure martingale on $G(n,p)$. Fix an ordering of the $t = \binom{n}{2}$ edges of K_n. For $1 \leq i \leq t$, let Z_i be the indicator random variable of whether the ith edge is present. Let $A = f(Z_1, \ldots, Z_t)$ be a fixed graph-theoretic function defined on $G \in G(n,p)$, such as the chromatic number of G. Then the Doob martingale $(X_i : 0 \leq i \leq t)$ in this case is the *edge-exposure martingale*. In the beginning we have no information about G. We are given more information over time; at time i we know which of the first i edges are in G.

The *vertex-exposure martingale* is similar, but one vertex is exposed at a time. Fix a linear ordering of $[n]$. Let Z_i be an $(n-i)$-sequence of 0's and 1's, indicating whether there is an edge between vertex i and vertices $j > i$. Then the Doob martingale $(X_i : 0 \leq i \leq t)$ in this case is called the vertex-exposure martingale.

The sequence $(X_i : 0 \leq i \leq t)$ satisfies the *c-Lipschitz condition*, where $c > 0$ is an integer, if for all $0 \leq i \leq t-1$

$$|X_{i+1} - X_i| \leq c.$$

Put another way, the differences between successive random variables are not too large. Similarly, a function $f(Z_1, \ldots, Z_t)$ is *c-Lipschitz* if changing the value of any one coordinate of f causes f to change by at most a constant c in absolute value. The two Lipschitz conditions for sequences of random variables and functions are bridged by the following theorem (for a proof, see Section 12.5.1 of [**165**]).

THEOREM 3.18. *If f is c-Lipschitz and Z_i is independent of $Z_{i+1}, \ldots, Z_n$ conditioned on $Z_1, \ldots, Z_{i-1}$, then the Doob martingale $(X_i : 0 \leq i \leq t)$ of f with respect to Z_i is c-Lipschitz.*

The following important concentration result for martingales was proved independently by Azuma [**16**] and Hoeffding [**134**].

THEOREM 3.19 (Azuma-Hoeffding inequality). *If $(X_i : 0 \leq i \leq t)$ is a martingale satisfying the c-Lipschitz condition, then for all real $\lambda > 0$,*

$$\mathbb{P}(|X_t - X_0| \geq \lambda\sqrt{t}) \leq 2\exp\left(-\frac{\lambda^2}{2c^2}\right). \tag{3.16}$$

In other words, if the c-Lipschitz condition holds, then the values of X_t are *concentrated around* X_0. More precisely, if $t \to \infty$ and $\lambda \to \infty$ very slowly as a function of t, then with probability $1 - 2\exp(-\frac{\lambda^2}{2})$, X_t lies in an interval around X_0 of length $2\lambda\sqrt{t}$. This method of proving concentration for random variables has been called the *method of bounded differences* (see [**160**]).

Proof of Theorem 3.19. It is sufficient to prove the inequality

$$\mathbb{P}(X_t \geq X_0 + \lambda\sqrt{t}) \leq \exp\left(-\frac{\lambda^2}{2c^2}\right). \tag{3.17}$$

Then to prove (3.16), apply (3.17) to the martingale $(-X_i : 0 \leq i \leq t)$.

For $1 \leq i \leq t$, define $Y_i = X_i - X_{i-1}$. Then $|Y_i| \leq c$. Further, for all $1 \leq i \leq t$,

$$\mathbb{E}(Y_i|X_0, \dots, X_i - 1) = 0,$$

since $(X_i : 0 \leq i \leq t)$ is a martingale. By the convexity of the function $f(x) = e^{nx}$ for $n > 0$, we have that for all $1 \leq i \leq t$

$$\begin{aligned}
&\mathbb{E}\left(\exp(nY_i)\,|X_0, \dots, X_{i-1}\right)\\
&\leq \mathbb{E}\left(\frac{1}{2c}\left(e^{nc} - e^{-nc}\right)Y_i + \frac{1}{2}\left(e^{nc} + e^{-nc}\right)|X_0, \dots, X_{i-1}\right)\\
&= \frac{1}{2}\left(e^{nc} - e^{-nc}\right)\\
&\leq \exp(n^2c^2/2),
\end{aligned} \tag{3.18}$$

where the first equality follows by the linearity of expectation, and the second equality follows by considering the Maclaurin series for e^{nc} and e^{-nc}.

We therefore have that

$$\begin{aligned}
&\mathbb{E}\left(\exp(n(X_t - X_0))\right)\\
&= \mathbb{E}\left(\prod_{i=1}^{t} e^{nY_i}\right)\\
&= \mathbb{E}\left(\left(\prod_{i=1}^{t-1} e^{nY_i}\right)\mathbb{E}\left(\exp(nY_t)\,|X_0, \dots, X_{t-1}\right)\right)\\
&\leq \mathbb{E}\left(\prod_{i=1}^{t-1} e^{nY_i}\right)\exp(n^2c^2/2)\\
&\leq \exp(n^2c^2t/2),
\end{aligned} \tag{3.19}$$

where the first inequality follows from (3.18), and the last inequality is obtained by induction. Let $n = \lambda/(c^2\sqrt{t})$. Then,

$$\begin{aligned}
\mathbb{P}(X_t \geq X_0 + \lambda\sqrt{t}) &\leq \mathbb{P}(\exp(n(X_t - X_0)) \geq \exp(n\lambda\sqrt{t}))\\
&\leq \exp(-n\lambda\sqrt{t})\mathbb{E}\left(\exp(n(X_t - X_0))\right)\\
&\leq \exp(-n\lambda\sqrt{t})\exp(n^2c^2t/2)\\
&= \exp\left(-\frac{\lambda^2}{2c^2}\right),
\end{aligned}$$

where the second inequality follows from Markov's inequality, and the third inequality is implied by (3.19). □

As an application of the method of bounded differences to random graphs, the following result was proved by Shamir and Spencer [**184**]. It tells us that the chromatic number of $G(n,p)$ concentrates around its expectation, but does not tell us what the expected value is!

THEOREM 3.20. *If $G \in G(n,p)$, then for all real $\lambda > 0$,*

$$\mathbb{P}(|\chi(G) - \mathbb{E}(\chi(G))| \geq \lambda\sqrt{n}) \leq 2\exp\left(-\frac{\lambda^2}{2}\right).$$

Proof. Consider the vertex-exposure martingale $(X_0, X_1, \ldots, X_n)$, with $f(Z_1, \ldots, Z_t) = \chi(G)$. Note that f is 1-Lipschitz, since changing any one coordinate can only change the chromatic number by 1. Hence, the vertex-exposure martingale is 1-Lipschitz. Now apply Theorems 3.18 and 3.19. □

The expected value of chromatic number for random graphs (with constant p) was computed by Bollobás [**30**]. We note that Theorem 3.20 was the first use of martingales within graph theory.

THEOREM 3.21 ([**30**]). *A.a.s. for $G \in G(n,p)$, we have that*

$$\mathbb{E}(\chi(G)) = (1 + o(1))\frac{n}{2\log_{\frac{1}{1-p}} n}.$$

Much more is known about the chromatic number of random graphs (see Chapter 11 of [**32**]). For example, Alon and Krivelevich [**11**] proved that for every constant $c > 0$ the chromatic number of $G(n,p)$ with $p = n^{-\frac{1}{2}-c}$ is a.a.s. concentrated on two values.

Exercises

(1) (a) Show that $R(3) = 6$ and $R(4) = 18$. As mentioned in Section 3.2, it is an open problem to determine $R(n)$, where $n \geq 5$.
 (b) Prove Theorem 3.1. Much remains unknown about the asymptotics of Ramsey numbers. For instance, a conjecture of Erdős (and one of his \$100 prize problems) is that the following limit exists:

$$\lim_{n\to\infty} R(n)^{1/n}.$$

(2) (a) Prove Theorem 3.3 in the case $p \in (0,1)$ is fixed.
 (b) Find examples of 2-e.c. graphs for all orders $n \geq 9$.
 (c) [**41**] Prove that $K_3 \square K_3$ is the unique 2-e.c. graph of order 9.
 (d) Show that a 3-e.c. graph must have order at least 20. It is an open problem whether a 3-e.c. graph of order 20 exists.

(3) [**18**] A *Steiner triple system* is a 2-$(n, 3, 1)$ *design*; that is, a collection of 3-subsets called *blocks*, of an n-set whose elements are called *points*, such that each distinct pair of elements of this n-set is contained in a unique block. A *Kirkman triple system* is a Steiner triple system with the property that the blocks of it can be partitioned into sets, called *parallel classes*, so that each point of the design is contained in a unique block of each parallel class.
   (a) Find a Kirkman triple system of order 9.
   (b) Show that if a Kirkman triple system of order n exists, then $n \equiv 3 \pmod 6$.
   (c) Adapt the proof of Theorem 3.9 to generate regular n-e.c. graphs from Kirkman triple systems.

(4) Verify that the following properties hold a.a.s. for G in $G(n, p)$ with p a fixed real number in $(0, 1)$.
   (a) The graph G has diameter 2.
   (b) The graph G is not planar.
   (c) For a finite fixed H, the graph G contains H as an induced subgraph.

(5) Prove directly that $G \in G(n, \frac{1}{2})$ a.a.s. satisfies the quasi-random properties (Pi), $i = 1$, 4, and 5.

(6) (a) Verify that the graph $G(q, S, A)$ is an

$$\mathrm{SRG}(q^2, |S|(q-1), q-2+(|S|-1)(|S|-2), |S|(|S|-1)).$$

   (b) If q is odd and $|S| = \frac{q+1}{2}$, then show that a graph in $G(q, S, A)$ is quasi-random.

(7) For an integer $n \geq 1$, define the graph $G(n)$ to have vertices the n-sets of a fixed $2n$-set. Two vertices u and v are joined if and only if

$$|u \cap v| \equiv 0 \pmod 2.$$

Prove that the graph $G(n)$ is quasi-random.

(8) [**189**] Prove that for each $n \geq 1$, there is a tournament of order n with at least $n!2^{-(n-1)}$ directed Hamilton paths.

(9) Prove that for all real $\varepsilon > 0$, there are graphs G on $2n$ vertices with average degree d at least $(c \log 2)/\varepsilon^2$, where c is some constant, such that for every partition of $V(G)$ into two parts A and B of equal cardinality, the number of edges between X and Y is at least $(1-\varepsilon)\frac{dn}{2}$ and at most $(1+\varepsilon)\frac{dn}{2}$.

(10) (a) A set $S \subseteq V(G)$ is a *dominating set* if each vertex not in S is joined to some vertex of S. Let G be a graph with minimum degree $\delta > 1$. Using the methods of this chapter, show that G has a dominating

set with cardinality at most

$$n\left(\frac{1+\log(\delta+1)}{\delta+1}\right).$$

(b) Prove, using the methods of this chapter, that a graph with e edges contains a bipartite subgraph with at least $\frac{e}{2}$ edges.

(11) Prove Theorem 3.5.

(12) For $m \geq 3$ an integer, let X be the random variable counting the number of cycles of length at most m in $G(n,p)$. Show that

$$\mathbb{E}(X) = \sum_{i=3}^{m} \binom{n}{i} \frac{(i-1)!}{2} p^i.$$

(13) Define a sequence of graphs $(M_n : n \geq 3)$ inductively as follows. Let G_3 be C_5. Form $G_{n+1} = (V_{n+1}, E_{n+1})$ as follows. If V_n is the vertex set of G_n, then $V_{n+1} = V_n \cup V_n' \cup \{x\}$, where $V_n' = \{u' : u \in V_n\}$ is disjoint from V_n and x is not in $V_n \cup V_n'$. Define E_{n+1} so that we keep edges in V_n, V_n' is independent, and $u'v$ is an edge if and only if $uv \in E_n$, and x is joined to each vertex of V_n'. The graphs M_n are the *Mycielski graphs*. Prove that M_n is triangle-free with $\chi(M_n) = n+1$.

(14) Show that for p fixed, w.e.p. every pair of distinct vertices of $G \in G(n,p)$ has

$$p^2 n + O(\sqrt{p^2 n}\log n) = (1+o(1))p^2 n$$

many common neighbours.

(15) Prove that for $p \in (0,1)$ fixed, w.e.p. all independent sets of $G \in G(n,p)$ have less than $n^{1/3}$ vertices.

(16) [**100**] Prove that a.a.s. $G \in G(n, \frac{1}{2})$ has a trivial automorphism group.

(17) The celebrated *Lovász local lemma* [**97**] states that if $A_1, A_2, \ldots, A_n$ are events in a probability space such that each A_i occurs with probability at most p, each A_i is independent of all the other A_j with at most d exceptions, and $ep(d+1) \leq 1$, then there is a non-zero probability that none of the events occur. Using the Lovász local lemma, prove that if

$$e\left(\binom{n}{2}\binom{r}{n-2}+1\right)2^{1-\binom{n}{2}} < 1,$$

then $R(n) > r$.

(18) Prove that cycles, trees, and cliques are balanced.

(19) A graph G is *(p,α)-jumbled* if p and α are real numbers satisfying $0 < p < 1 \leq \alpha$ such that every set $U \subseteq V(G)$ satisfies

$$\left| |E(G \restriction U)| - p\binom{|U|}{2} \right| \leq \alpha |U|.$$

Show that a.a.s. $G(n,p)$ is $O(\sqrt{np})$-jumbled.

(20) (a) Prove that for every $\varepsilon > 0$ and $p = (1-\varepsilon)(\log n)n^{-1}$, a.a.s. $G \in G(n,p)$ has an isolated vertex.

(b) Suppose that $p = \frac{\log n + c}{n}$, where $c > 0$ is a constant that does not depend on n. Prove that

$$\lim_{n\to\infty} \mathbb{P}(G \in G(n,p) \text{ has no isolated vertex}) = e^{-e^{-c}}.$$

(21) Prove that the property $\omega(G) \geq 4$ has $n^{-2/3}$ as a threshold function.

(22) For a positive integer k, suppose that $n^{-1-\frac{1}{k}} = o(p)$ and $p = o(n^{-1-\frac{1}{k-1}})$. Prove that the following properties hold a.a.s. for $G \in G(n,p)$.

(a) There are no connected components in G with at least $k+2$ vertices.

(b) The graph G is acyclic.

(c) For all integers r, $k \geq 1$, every tree T of order at most $k+1$ appears as a connected component at least r times.

(23) (a) Prove that for X, Y random variables on a common probability space,

$$\mathbb{E}[\mathbb{E}[X|Y]] = \mathbb{E}[X].$$

(b) Prove the tower property of conditional expectation, Theorem 3.16.

(c) Prove Lemma 3.17.

(24) [**30**] For $k > 0$ an integer, let Y be the maximal size of a family of edge disjoint cliques of order k in $G(n, \frac{1}{2})$.

By using the edge-exposure martingale, prove that for $G \in G(n, \frac{1}{2})$ and $k > 1$ an integer, there is a positive constant c such that

$$\mathbb{P}(\omega(G) < k) < \exp\left(-(c + o(1))\frac{n^2}{\log^8 n}\right).$$

Chapter 4

Models for the Web Graph

There is no branch of mathematics, however abstract, which may not some day be applied to phenomena of the real world.

–Nikolai Lobachevsky

All models are wrong, but some models are useful.

–G.E.P. Box

4.1. Introduction

While graph theory is usually studied in the abstract and is a fascinating subject in its own right, one of its long-recognized applications is to model networks in the real world. The web graph W is one such real-world network, and is our focus. In modelling the web, the many parameters and complex phenomena at work in its evolution are distilled into a simplified picture. Good models for W usually have only a few parameters, and must capture salient features of W. The balance between simplicity of design and the difficulties of analyzing the model makes the subject of modelling W challenging.

In this chapter we will discuss some of the key models for the web graph W. Our approach will be to consider *stochastic models*: that is, models where graphs are generated over an infinite sequence of discrete time-steps via certain probabilistic rules. Early models of W were often posed and analyzed in a non-rigorous fashion. The present challenge is to design mathematically rigorous models that simulate one or more of the features observed in W (as described in Chapter 2), and can be rigorously analyzed. As we will see, this challenge has been met by an increasing number of rigorous models. To analyze these models, the techniques from Chapter 3 come into play, as well as new ones that we describe as they are needed.

What features make a good web graph model? There is no definitive answer to this question, only a consensus based on observed properties of W. The following is a partial list of desirable properties that graphs generated by a web graph model should possess, based on the observed properties of W described in Chapter 2.

(1) *On-line property.* Graphs generated by the model change with time, so both the number of vertices and the number of edges change with time.

(2) *Power law degree distribution.* A.a.s. the degree distribution of graphs generated by the model follows a power law with exponent $\beta > 2$.

(3) *Small world property.* The model a.a.s. generates sparse graphs with "low" diameter and average distance. For example, the diameter should be a.a.s. approximately $\log t$ if there are t vertices, while the average distance should be approximately $\log \log t$.

Additional, but less frequently studied, desirable properties for a model of W include higher values of the clustering coefficient when compared to a random graph of similar size and order, a larger number of bipartite cliques when compared to a random graph of similar size and order, and sparse cuts. For simplicity, we will focus on properties (1), (2), and (3) above, but we will mention some of these other properties along the way as we describe various models.

Pioneering work on random graphs was first done by Erdős and Rényi [**98, 99**], as discussed in Chapter 3. We begin by recalling the random graph $G(n,p)$. We are given n vertices and a fixed real number parameter $p \in (0,1)$. For each of the $\binom{n}{2}$ many distinct pairs of vertices, add an edge between them independently with probability p. The probability space $G(n,p)$, in a certain sense, is static or *off-line*: the number of vertices is fixed. Although usually n is taken as very large, and the number of edges is viewed as being variable over time, the number of vertices in $G(n,p)$ is off-line. Few techniques or models were available before the late 1990's for on-line random graph models. Further, from Theorem 3.11 the degrees of vertices are binomially distributed. Hence, based on items (1) and (2) above, $G(n,p)$ is not appropriate as a model of the web graph W (after all, the study of random graphs predates the inception of the internet by several decades). Nevertheless, random graphs supply the mathematical subtext for these new models, and many of the techniques used to analyze them are also useful for models of W.

Over the last decade, a large number of rigorous models for the web graph W have been proposed. Such models deepen our understanding of the generative mechanisms driving the evolution of W, and provide insight into superficially unrelated properties observed in the web.

We make two caveats. First, we focus on a handful of the most influential models, and on the properties (1), (2), and (3) described above. As our goal is an introduction to models of W, we do not claim to survey all the models in the literature. Models of W (and many other self-organizing networks) have been analyzed with regards to several other properties, such as their eigenvalues [**78, 79, 161**], vulnerability to attack [**36**], orders of their connected components [**75**], and spread of viruses and worms on W [**25**]. See Chapter 7 for a discussion of some of these topics. The second caveat is that, since this is a mathematics text, we focus only on rigorous models and analysis.

In Section 4.2.1 we consider preferential attachment models for W, and give a rigorous analysis of the degree distribution of one of these models. Other models are described, including the copying model (Section 4.2.5), growth-deletion models (Section 4.2.6), geometric models (Section 4.2.7), and an off-line model (Section 4.2.8). We finish with a description of some challenges faced in future modelling of W.

4.2. On-Line Web Graph Models

As discussed in the previous section, our emphasis is on *on-line* web graph models: that is, models whose vertex set increases in cardinality over time. This approach is the most desirable one for models of W, which is, after all, a dynamic graph. A central idea in all models is to consider both approximate results and asymptotic behaviour. The rationale for both is that W is a massive graph with a large number of vertices and edges, and on average, small changes make little difference in the structure of the overall graph. Random asymptotic techniques are therefore most suitable for analyzing these models.

To simplify notation, we supply the following framework for all the on-line models we present. The model will always possess a finite set of real number parameters (the fewer the better), and has a fixed finite graph H as an additional parameter. The model generates by some random graph process a sequence of finite graphs G_t indexed by $(t : t \in \mathbb{N})$. Unless otherwise stated, for all $t \in \mathbb{N}$, we have that

(1) $G_0 \cong H$;

(2) G_t is an induced subgraph of G_{t+1};

(3) $|V(G_{t+1})| = |V(G_t)| + 1$.

In all the on-line models we consider, the graphs G_t are defined inductively. In the inductive step, the unique vertex in $V(G_{t+1})\backslash V(G_t)$ is referred to as the *new vertex*, written v_{t+1}, and the vertices of $V(G_t)$ are the *existing vertices*. We note that the choice of H usually has no effect on the value of the power law exponent β, while the choice of real number parameters does generally affect β. Further, the number of edges in $E(G_{t+1})\backslash E(G_t)$ is usually a constant parameter. However, this is not always the case: the numbers of edges and vertices in G_t may be random variables.

For $k, t \geq 0$ integers, define $N_{k,t}$ to be the number of vertices of degree k at time t. Then $N_{k,t}$ itself is a random variable. Since the number of vertices in G_t is usually approximately t, most authors study the proportion $\frac{N_{k,t}}{t}$. The most common technique for proving that $\frac{N_{k,t}}{t}$ follows a power law involves first computing $\mathbb{E}\left(\frac{N_{k,t}}{t}\right)$, and then proving that $\frac{N_{k,t}}{t}$ does not deviate too far from $\mathbb{E}\left(\frac{N_{k,t}}{t}\right)$. Authors prove that the random variable *concentrates* on its expected value. As we will see, both the computation of the expected value of $\frac{N_{k,t}}{t}$ and deriving concentration around the expected value are non-trivial problems for models of W.

4.2.1. Preferential attachment models. Arguably the most important web graph models are ones incorporating some form of preferential attachment. The first evolving graph model explicitly designed to model W was given by Barabási and Albert [**21**]. The idea behind their model is an intuitively pleasing one: new vertices are more likely to join to existing vertices with high degree. In a slogan, *the rich get richer.* This model is now referred to as an example of a *preferential attachment* (or *PA*) *model.* As we will discuss in detail, PA models generate graphs with properties quite different from those of $G(n, p)$. See Figures 4.1 and 4.2 to compare and contrast graphs generated by these two models.

Barabási and Albert gave a heuristic description and analysis of their PA model (using mean field theory from physics), and concluded that it generates graphs whose in-degree distribution follows a power law with exponent $\beta = 3$. Although their proof was not rigorous, their important work set the stage for most of the mathematics regarding the modelling of W to come.

4.2.2. The LCD PA model. The first rigorous analysis of a PA model was given in Bollobás, Riordan, Spencer, and Tusnády [**34**]. Their model is called the *Linearized Chord Diagram* or *LCD* model, since an equivalent

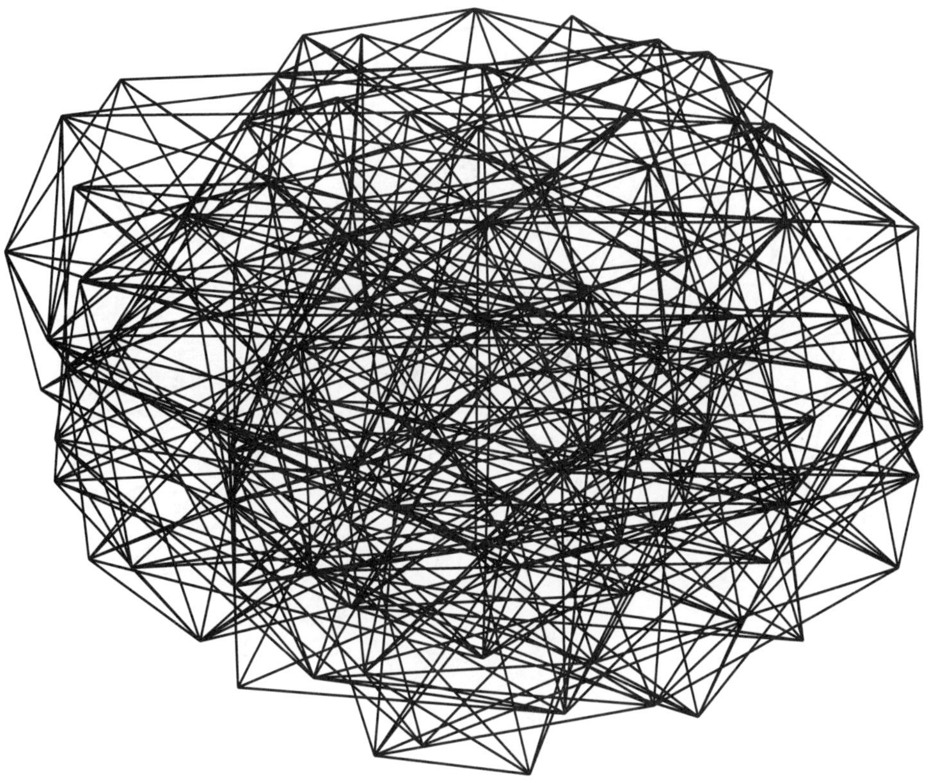

Figure 4.1. A graph with 100 vertices and edges drawn with probability $\frac{1}{2}$.

formulation of the model is via random pairings on a fixed finite set of integers, as we will describe below. The sole parameter of this model is a positive integer m, where H is a copy of K_1 with a single loop. We first describe the model in the case $m = 1$. To form G_t from G_{t-1} add a single edge from v_t to v_i, where the vertex v_i is chosen at random from the existing vertices, with

$$\mathbb{P}(i = s) = \begin{cases} \frac{\deg_{G_{t-1}}(v_s)}{2t-1} & \text{if } 1 \le s \le t-1, \\ \frac{1}{2t-1} & \text{if } s = t. \end{cases} \tag{4.1}$$

Note that (4.1) gives a higher probability for an existing vertex to acquire a new edge if it has high degree. The graph G_t contains no non-trivial cycles, although self-loops are allowed. A similar version of this model was previously studied (in a different context) as *random recursive trees*; see [**34**] for further discussion.

If $m > 1$, then define the process $(G_t^m : t \geq 0)$ by first generating a sequence $(G_t : t \in \mathbb{N})$ of graphs using the case $m = 1$ on a sequence of

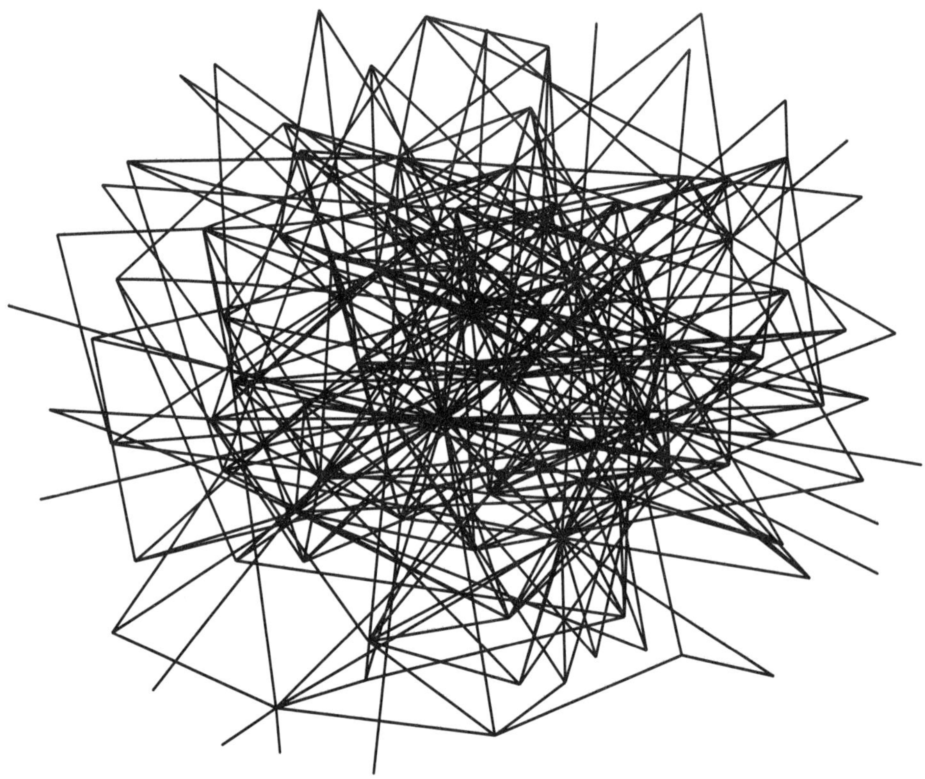

Figure 4.2. A graph generated by the preferential attachment model with 100 vertices and average degree close to the average degree of the graph in Figure 4.1.

vertices $(v_i' : i \in \mathbb{N} \setminus \{0\})$. The graph G_t^m is formed from G_{mt} by identifying the vertices $v'_{(i-1)m+1}, \ldots, v'_{im}$ to form v_i.

To analyze the model, the authors use the following geometric notion. Define a *linearized chord diagram* or *LCD* of order t to be a partition of $[2t]$ into n sets of distinct pairs. Then there are $\frac{(2t)!}{t!2^t}$-many LCD's on $[2t]$ (see Exercise 3).

We may identify an LCD with t-many semi-circular chords between $2n$ distinct points on the x-axis paired off in the upper half-plane of $\mathbb{R}^2$. Each chord has a left and right endpoint. For a fixed LCD L we may form an undirected graph $G(L)$ of order t as follows. To form vertex v_1, identify all endpoints up to and including the first right endpoint of all the chords. Proceed inductively to define the remaining vertices, so that the $(k+1)$th vertex is formed by identifying all endpoints up to and including the first right endpoint after vertex v_k. The chords may be viewed as (multiple) edges among the vertices. For an example, see Figure 4.3.

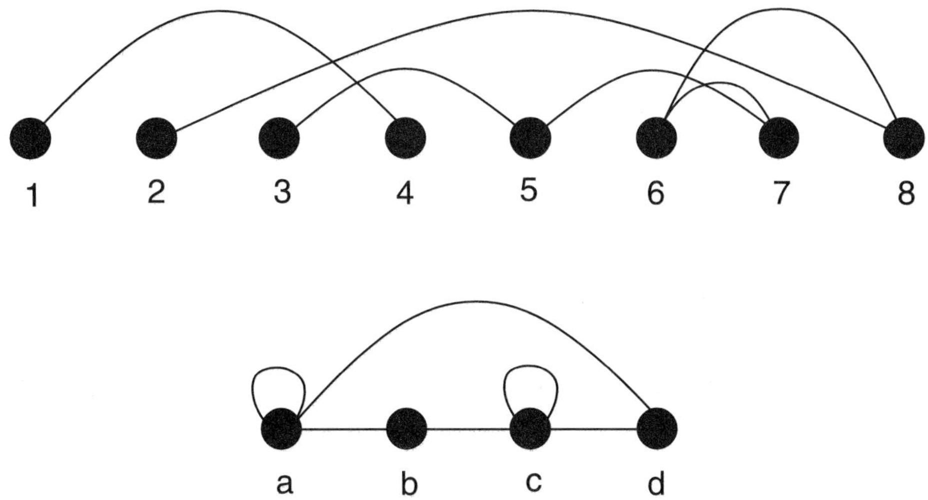

Figure 4.3. An LCD with the corresponding graph. The vertex a is identified with $1, 2, 3, 4$, the vertex b with 5, the vertex c with $6, 7$, and d with 8.

The connection between LCD's and the LCD model comes from the following theorem (whose proof is an exercise).

THEOREM 4.1. *Let $m = 1$. Suppose that an LCD L is chosen u.a.r. from the $\frac{(2t)!}{t!2^t}$ many LCD's on $[2t]$. Then the probability that vertices v_i and v_s are joined is given by (4.1).*

Bollobás et al. prove the following theorem.

THEOREM 4.2 ([**34**]). *In the LCD model, fix m a positive integer, and fix $\varepsilon > 0$. For k a non-negative integer, define*

$$\alpha_{m,k} = \frac{2m(m+1)}{(k+m)(k+m+1)(k+m+2)}.$$

Then a.a.s. for all k satisfying $0 \leq k \leq t^{1/15}$,

$$(1-\varepsilon)\alpha_{m,k} \leq \frac{N_{k,t}}{t} \leq (1+\varepsilon)\alpha_{m,k}. \tag{4.2}$$

Theorem 4.2 demonstrates that for large t, with high probability the degree distribution of graphs in the LCD model follows a power law with exponent $\beta = 3$ (formally justifying the conclusions derived in [**21**]). The reader will note that Theorem 4.2 is stated as a concentration result for degrees in the range $0 \leq k \leq t^{1/15}$; as remarked in [**34**], this may be extended to degrees $k > t^{1/15}$. An important observation is that the power law exponent $\beta = 3$ is independent of the choice of m. While $\beta = 3$ is in the correct range to model W, this restrictiveness of the power law exponent can be viewed as a drawback of the LCD model. In Chapter 6, we explore how infinite limit graphs play a role in distinguishing different values of m.

The non-trivial part of the proof of Theorem 4.2 involves estimating $\mathbb{E}(N_{k,t})$ to be $\alpha_{m,k}t$. Rather than give a full proof of Theorem 4.2, we will prove a similar result for a closely related, but more easily stated PA model in Section 4.2.3.

THEOREM 4.3. *For all k satisfying $0 \leq k \leq t^{1/15}$, the sequence $(N_{k,t} : t \in \mathbb{N})$ converges in probability to $\mathbb{E}(N_{k,t})$.*

Using Theorem 4.3 and assuming that $\mathbb{E}(N_{k,t}) \sim \alpha_{m,k}t$ in the range $0 \leq k \leq t^{1/15}$, (4.2) follows. We give a proof of the concentration result in the case $m = 1$ (for the general case, see [**34**]), using the method of bounded differences as described in Chapter 3.

Proof of Theorem 4.3 *if* $m = 1$. Using notation from Chapter 3, consider the Doob martingale $(X_i : 0 \leq i \leq t)$ defined by $X_0 = \mathbb{E}(A)$ and let $X_i = \mathbb{E}[A|Z_1, \ldots, Z_i]$, where $A = N_{k,t}$ and $Z_i = G_i^m$ (we consider the graphs from earlier time-steps as random variables). Since a new vertex can affect the degrees of at most two existing vertices, we claim that for $1 \leq i \leq t$,

$$|X_i - X_{i-1}| \leq 2. \tag{4.3}$$

The reason for this is that whether at time-step i the vertex v_i is joined to vertices v_r or v_s does not affect the degrees at later times of vertices v_u, with u distinct from r and s. The joint distribution of all other degrees is the same in either case. As vertices with a given degree are counted, however the degrees of v_r and v_s change in G_t^m, the result is that (4.3) holds. (For an alternate explanation of (4.3) using so-called "half-edges" see the proof of Theorem 1 in [**34**].) By Theorems 3.18 and 3.19 of Chapter 3, we have that with $\lambda = \sqrt{\log t}$

$$\mathbb{P}(|X_t - X_0| \geq \sqrt{\log t}\sqrt{t}) < 2t^{-\frac{1}{8}} = o(1).$$

In particular, X_t converges to X_0 in probability. □

Bollobás and Riordan [**37**] prove the following non-trivial result (whose proof is omitted) which computes the diameter of G_t^m, verifying that the LCD model generates small world graphs. Hence, the LCD model generates graphs satisfying all three of the properties from Section 4.1, and so it is our first satisfactory model for the web graph.

THEOREM 4.4 ([**37**]). *Fix an integer $m \geq 2$ and a positive real number ε. Then a.a.s. G_t^m satisfies*

$$(1-\varepsilon)\frac{\log t}{\log\log t} \leq \mathrm{diam}(G_t^m) \leq (1+\varepsilon)\frac{\log t}{\log\log t}.$$

As with Theorem 4.2, the result of Theorem 4.4 is independent of m. The case where $m = 1$ is not included as the upper bound does not hold there in general.

4.2.3. Rigorous analysis of a PA model. We describe a PA model $\mathcal{G}(m)$ [**138**] which has a simpler description than the LCD model, without the need for the identification of vertices. In addition, the m edges added at time $t+1$ are added *independently*. In the case $m = 1$, this approach coincides exactly with the LCD model. A model analogous to $\mathcal{G}(m)$ for general m was given in [**139**]. Although the independence is less realistic as a model for web page creation, it makes the mathematical analysis easier (however, the proof of the power law is technical).

The sole parameter of $\mathcal{G}(m)$ is a positive integer m. The initial graph G_0 is a fixed finite, connected graph with t_0 vertices and e_0 edges. To form G_{t+1}, we add a single vertex v_{t+1} to G_t. The vertex v_{t+1} is joined via m edges to vertices $w_{t+1,i}$, where $1 \leq i \leq m$, via preferential attachment. More precisely, the probability that $w_{t+1,i}$ is w equals

$$\frac{\deg_{G_t}(w)}{\sum_{u \in V(G_t)} \deg_{G_t}(u)}$$

independently for each $1 \leq i \leq m$. Note that $|V(G_t)| = t + t_0$ and $|E(G_t)| = mt + e_0$. Hence,

$$\mathbb{P}(w_{t+1,i} = w) = \frac{\deg_{G_t}(w)}{c_t t} \tag{4.4}$$

where $c_t = \frac{2(mt+1)}{t}$. Unlike the LCD model, there are no loops in the graphs G_t, but possibly multiple edges. See Figure 4.4 for an example.

The rest of the section is devoted to a rigorous proof of the power law degree distribution for the PA model $\mathcal{G}(m)$. The proof is long, and requires some care. We follow the proof set out in [**138**]. We note that the proof of power law degree distributions usually consist of two parts.

(1) Derive an asymptotic expression for $\mathbb{E}(N_{k,t})$ via a recurrence relation. Auxiliary lemmas concerning convergence of real sequences usually are required.
(2) Show that $N_{k,t}$ concentrates around $\mathbb{E}(N_{k,t})$. This is accomplished using either martingales (as in the proof of Theorem 4.3 in the previous section), or variance (which will be our approach in this section).

We now rigorously analyze the asymptotic behaviour of the proportion $\frac{N_{k,t}}{t}$ and its expected value $\mathbb{E}\left(\frac{N_{k,t}}{t}\right)$. Note that $|V(G_t)| = t + t_0$, but we

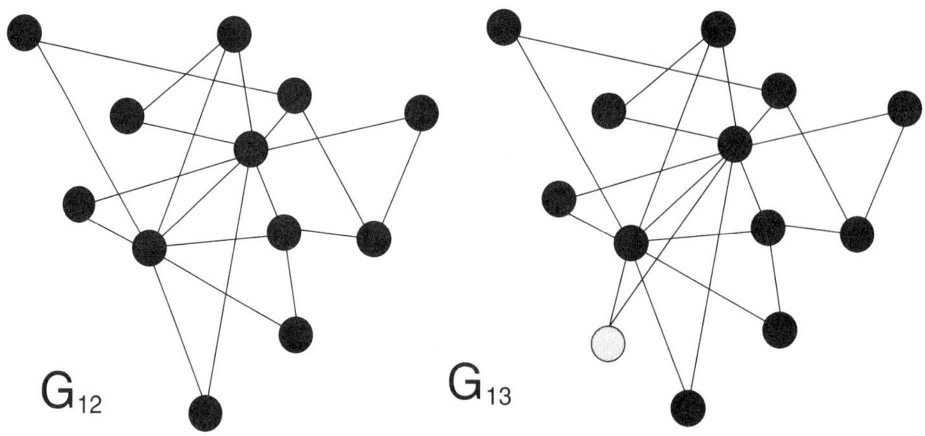

Figure 4.4. The graphs G_{12} and G_{13} in a simulation of $\mathcal{G}(2)$. The new vertex is grey, and is more likely to join to existing vertices with high degree.

divide by t rather than $t + t_0$. Since the results we present are asymptotic, this simplification causes no loss of generality.

We first prove the following lemma from [**138**] about sequences of real numbers, which will be useful for our purpose.

LEMMA 4.5. *For $t \in \mathbb{N}$, let x_t, y_t, η_t, and r_t be real numbers satisfying*

$$x_{t+1} - x_t = \eta_{t+1}(y_t - x_t) + r_{t+1} \tag{4.5}$$

and

(1) $\lim_{t\to\infty} y_t = x$;
(2) *For all t, $\eta_t > 0$, and for all sufficiently large t, $\eta_t < 1$;*
(3) $\sum_{t=1}^{\infty} \eta_t = \infty$;
(4) $\lim_{t\to\infty} \frac{r_t}{\eta_t} = 0$.

Then $\lim_{t\to\infty} x_t = x$.

Proof. By replacing y_t with $y_t + \frac{r_t}{\eta_t}$, in view of item (4) we may assume without loss of generality that $r_t = 0$ for all t. Hence, (4.5) simplifies to

$$x_{t+1} - x_t = \eta_{t+1}(y_t - x_t), \tag{4.6}$$

and so

$$x_{t+1} = x_t(1 - \eta_{t+1}) + \eta_{t+1} y_t. \tag{4.7}$$

Fix a real number $\varepsilon > 0$. By (1), we may choose a positive integer N so that for all $t \geq N$,

$$|y_t - x| < \frac{\varepsilon}{2}. \tag{4.8}$$

Hence, if $x_t > x - \varepsilon$ for some $t > N$, then by (2) and (4.7), $x_{t+1} > x - \varepsilon$ for some $t > N$. By a similar argument, if $x_t < x + \varepsilon$ for some $t > N$, then $x_{t+1} < x + \varepsilon$ for some $t > N$.

If $x_t < x - \varepsilon$ for all $t \geq N$, then by (4.6), (4.8), and item (3) we have that

$$\sum_{t=N}^{\infty} (x_{t+1} - x_t) \geq \sum_{t=N}^{\infty} \eta_{t+1}\frac{\varepsilon}{2} = \infty. \tag{4.9}$$

But (4.9) implies that $\lim_{t\to\infty} x_t = \infty$, which contradicts that $x_t < x - \varepsilon$ for all $t \geq N$. Similarly if $x_t > x + \varepsilon$ for all $t \geq N$, we derive the contradiction that

$$\sum_{t=N}^{\infty} (x_t - x_{t+1}) = \infty.$$

Hence, for t sufficiently large, $|x_t - x| \leq \varepsilon$. □

Using Lemma 4.5 and an analysis of the first moment of $\frac{N_{k,t}}{t}$, we have the following theorem, first proved in [**138**].

THEOREM 4.6. *For $k \geq m \geq 1$, define the constant*

$$\beta = \frac{2m(m+1)}{k(k+1)(k+2)}.$$

In the model $\mathcal{G}(m)$ the sequence $\mathbb{E}\left(\frac{N_{k,t}}{t}\right)$ converges to β as t tends to ∞.

Proof. For a fixed d with $0 \leq d \leq m$, by (4.4) the probability that $w_{t+1,i} = w$ for exactly d values of i is

$$\begin{aligned} &\binom{m}{d}\left(1 - \frac{\deg_{G_t}(w)}{c_t t}\right)^{m-d}\left(\frac{\deg_{G_t}(w)}{c_t t}\right)^d \\ &= \binom{m}{d}\left(\frac{c_t t - \deg_{G_t}(w)}{c_t t}\right)^{m-d}\left(\frac{\deg_{G_t}(w)}{c_t t}\right)^d. \end{aligned} \tag{4.10}$$

From (4.10), conditional on $N_{k-d,t}$, the expected number of vertices of degree k in G_{t+1} and degree $k-d$ in G_t is

$$N_{k-d,t}\binom{m}{d}\left(\frac{c_t t - k + d}{c_t t}\right)^{m-d}\left(\frac{k-d}{c_t t}\right)^d. \tag{4.11}$$

Let $\mathcal{G}_t$ be the graphs $\{G_0, \ldots, G_t\}$. Let I_m be defined by

$$I_m(k) = \begin{cases} 1 & \text{if } k = m, \\ 0 & \text{else.} \end{cases}$$

(The value $I_m(k)$ is often called a *Kronecker delta.*) Since at time-step $t+1$ exactly one new vertex is added with degree m, we have by (4.11) that

$$\mathbb{E}\left(N_{k,t+1}|\mathcal{G}_t\right) = \sum_{d=0}^{m} N_{k-d,t}\binom{m}{d}\left(\frac{c_t t - k + d}{c_t t}\right)^{m-d}\left(\frac{k-d}{c_t t}\right)^{d} + I_m(k).$$

Hence,

$$\begin{aligned} &\mathbb{E}\left(\frac{N_{k,t+1}}{t+1}|\mathcal{G}_t\right) \\ (4.12)\qquad &= \frac{t}{t+1}\sum_{d=0}^{m}\frac{N_{k-d,t}}{t}\binom{m}{d}\left(\frac{c_t t - k + d}{c_t t}\right)^{m-d}\left(\frac{k-d}{c_t t}\right)^{d} \\ &\quad + I_m(k)\frac{1}{t+1}. \end{aligned}$$

Equation (4.12) is sometimes referred to as the *master equation* (a term emanating from physics) describing the conditional expected value of $\frac{N_{k,t+1}}{t+1}$. Hence, $\mathbb{E}\left(\frac{N_{k,t+1}}{t+1} - \frac{N_{k,t}}{t}|\mathcal{G}_t\right)$ equals

(4.13)

$$\begin{aligned} \frac{1}{(t+1)(c_t t)^m}\Bigg(&\frac{N_{k,t}}{t}(t(c_t t - k)^m - (t+1)(c_t t)^m) \\ &+ \frac{N_{k-1,t}}{t} m t (k-1)(c_t t - k + 1)^{m-1} \\ &+ \sum_{d=2}^{m}\frac{N_{k-d,t}}{t}\binom{m}{d} t(k-d)^d (c_t t - k + d)^{m-d} + I_m(k)(c_t t)^m\Bigg). \end{aligned}$$

Upon simplification, (4.13) equals

$$\begin{aligned} \frac{1}{(t+1)c_t^m t^m}\Bigg(&t^m m c_t^{m-1}\left(\frac{N_{k-1,t}}{t}(k-1) + \frac{c_t}{m} I_m(k) - \frac{N_{k,t}}{t}\left(k + \frac{c_t}{m}\right)\right) \\ &+ m c_t^{m-1}\sum_{j=1}^{m} t^{m-j}\sum_{d=0}^{j} C_{k,j,d,t}\frac{N_{k-d,t}}{t}\Bigg), \end{aligned}$$

where $C_{k,j,d,t}$ is a function of k, j, d, t that is bounded in t.

Hence, $\mathbb{E}\left(\frac{N_{k,t+1}}{t+1}|\mathcal{G}_t\right) - \frac{N_{k,t}}{t}$ equals

(4.14)

$$\frac{1}{(t+1)}\left(\frac{m}{c_t}\right)\left(\left(k+\frac{c_t}{m}\right)\left(\frac{k-1}{k+\frac{c_t}{m}}\frac{N_{k-1,t}}{t} + \frac{c_t/m}{k+c_t/m}I_m(k) - \frac{N_{k,t}}{t}\right)\right.$$
$$\left. + \sum_{j=1}^{m} t^{-j} \sum_{d=0}^{j} C_{k,j,d,t}\frac{N_{k-d,t}}{t}\right).$$

By the linearity of expectation and property (3.15) of conditional expectations, $\mathbb{E}\left(\frac{N_{k,t+1}}{t+1}\right) - \mathbb{E}\left(\frac{N_{k,t}}{t}\right)$ equals

$$\frac{1}{(t+1)}\left(\frac{m}{c_t}\right)\left(\left(k+\frac{c_t}{m}\right)\left(\left(\frac{k-1}{k+\frac{c_t}{m}}\right)\mathbb{E}\left(\frac{N_{k-1,t}}{t}\right) + \frac{c_t/m}{k+c_t/m}I_m(k)\right.\right.$$
$$\left.\left. -\mathbb{E}\left(\frac{N_{k,t}}{t}\right)\right) + \sum_{j=1}^{m} t^{-j}\sum_{d=0}^{j} C_{k,j,d,t}\mathbb{E}\left(\frac{N_{k-d,t}}{t}\right)\right).$$

There are three cases to consider: $k < m$, $k = m$, and $k > m$. If $k < m$, since each new vertex has degree m, and as degrees only increase, there are at most $V(G_0)$-many vertices of degree k. It follows that

$$\lim_{t\to\infty} \mathbb{E}\left(\frac{N_{k,t}}{t}\right) = 0.$$

Next consider the case $k = m$. Let

$$\begin{aligned}
\eta_{t+1} &= \frac{1}{(t+1)}\left(\frac{m}{c_t}\right)\left(m+\frac{c_t}{m}\right), \\
r_{t+1} &= \frac{1}{(t+1)}\left(\frac{m}{c_t}\right)\sum_{j=1}^{m} t^{-j}\sum_{d=0}^{j} C_{k,j,d,t}\mathbb{E}\left(\frac{N_{k-d,t}}{t}\right), \\
y_t &= \left(\frac{1}{m+\frac{c_t}{m}}\right)\left(\frac{c_t}{m}\right) + \left(\frac{m-1}{m+\frac{c_t}{m}}\right)\mathbb{E}\left(\frac{N_{m-1,t}}{t}\right), \text{ and} \\
x_t &= \mathbb{E}\left(\frac{N_{k,t}}{t}\right).
\end{aligned}$$

Now,

$$\begin{aligned}
\sum_{t=1}^{\infty}\frac{1}{(t+1)}\left(\frac{m}{c_t}\right)\left(k+\frac{c_t}{m}\right) &= \sum_{t=1}^{\infty}\frac{1}{(t+1)}\left(\frac{km}{c_t}+1\right) \\
&\geq \left(\frac{k}{3}+1\right)\sum_{t=1}^{\infty}\frac{1}{(t+1)} = \infty,
\end{aligned}$$

where the inequality follows since $(c_t : t \in \mathbb{N})$ is a decreasing sequence which is bounded above (and with limit) $2m$, and the final equality follows since the harmonic series diverges. As $C_{k,j,d,t}$ is bounded in t and $\left|\mathbb{E}\left(\frac{N_{k-d,t}}{t}\right)\right| \leq 1$ (since $N_{k-d,t} \leq t$), we have that for all k, m, j,

$$\lim_{t\to\infty} t^{-j} \sum_{d=0}^{j} C_{k,j,d,t}\mathbb{E}\left(\frac{N_{k-d,t}}{t}\right) = 0.$$

Hence,

$$\lim_{t\to\infty} \frac{r_t}{\eta_t} = 0.$$

Then by the preceding discussion, the terms η_t, r_t, x_t, and y_t satisfy the hypotheses of Lemma 4.5, and so

$$\begin{aligned} \lim_{t\to\infty} x_t &= \lim_{t\to\infty} \mathbb{E}\left(\frac{N_{k,t}}{t}\right) \\ &= \lim_{t\to\infty} y_t \\ &= \frac{2}{m+2}. \end{aligned}$$

We finally consider the case $k > m$. We proceed by induction on $k > m$ using Lemma 4.5 with the same choices of η_t, r_t, x_t as in the case $k = m$, except that now

$$y_t = \left(\frac{k-1}{k+\frac{c_t}{m}}\right)\mathbb{E}\left(\frac{N_{k-1,t}}{t}\right).$$

In this case, $I_m(k) = 0$.

Using the induction hypothesis, we have that

$$\begin{aligned} \lim_{t\to\infty} \mathbb{E}\left(\frac{N_{k,t}}{t}\right) &= \frac{2}{m+2}\prod_{r=m+1}^{k}\frac{r-1}{r+2} \\ &= \frac{2m(m+1)}{k(k+1)(k+2)}, \end{aligned}$$

where the second equality follows since the product is telescoping. □

Theorem 4.6 implies that the expectation of $\frac{N_{k,t}}{t}$ follows a power law. However, the distribution of $\frac{N_{k,t}}{t}$ may be far from its mean. The concentration result for the degree distribution of graphs generated by $\mathcal{G}(m)$ is contained in the following theorem.

THEOREM 4.7. *In $\mathcal{G}(m)$, if $k \geq m$, then the proportions $\frac{N_{k,t}}{t}$ converge in probability to the limiting expectation*

$$\beta = \frac{2m(m+1)}{k(k+1)(k+2)}$$

as $t \to \infty$.

Proof. We analyze the variance of $\frac{N_{k,t}}{t}$, with the goal of proving that

$$\lim_{t\to\infty} \operatorname{Var}\left(\frac{N_{k,t}}{t}\right) = 0.$$

The result will then follow from Corollary 3.15 of Chapter 3. Recall (4.14) from the proof of Theorem 4.6 that $\mathbb{E}\left(\frac{N_{k,t+1}}{t+1}|\mathcal{G}_t\right) - \frac{N_{k,t}}{t}$ equals

$$\begin{aligned}(4.15)\qquad &\frac{1}{(t+1)}\left(\frac{m}{c_t}\right)\left(\left(k+\frac{c_t}{m}\right)\right.\\ &\times\left(\frac{k-1}{k+\frac{c_t}{m}}\frac{N_{k-1,t}}{t} + \frac{c_t/m}{k+c_t/m}I_m(k) - \frac{N_{k,t}}{t}\right)\\ &\left.+\sum_{j=1}^{m} t^{-j}\sum_{d=0}^{j} C_{k,j,d,t}\frac{N_{k-d,t}}{t}\right).\end{aligned}$$

Add $\frac{N_{k,t}}{t}$ to both sides of (4.15) and then square both sides of the equation. After grouping together $O(t^{-2})$ terms as t becomes large, we have that for $k > m$,

$$\begin{aligned}&\left|\mathbb{E}\left(\frac{N_{k,t+1}}{t+1}|\mathcal{G}_t\right)^2 - \left(\frac{N_{k,t}}{t}\right)^2 - \frac{2}{t+1}\left(\frac{m}{c_t}\right)\right.\\ &\left.\times\left(\left(k+\frac{c_t}{m}\right)\left(\left(\frac{k-1}{k+\frac{c_t}{m}}\right)\frac{N_{k-1,t}}{t}\frac{N_{k,t}}{t} - \left(\frac{N_{k,t}}{t}\right)^2\right)\right)\right| < C_k t^{-2},\end{aligned}$$

where C_k is a constant depending only on k. For $k = m$ we have that

$$\begin{aligned}&\left|\mathbb{E}\left(\frac{N_{m,t+1}}{t+1}|\mathcal{G}_t\right)^2 - \left(\frac{N_{m,t}}{t}\right)^2 - \frac{2}{t+1}\left(\frac{m}{c_t}\right)\right.\\ &\left.\times\left(\left(m+\frac{c_t}{m}\right)\left(\left(\frac{c_t/m}{m+\frac{c_t}{m}}\right)\frac{N_{m,t}}{t} - \left(\frac{N_{m,t}}{t}\right)^2\right)\right)\right| < C_m t^{-2},\end{aligned}$$

where C_m is a constant depending only on m.

When $k > m$ we have that

$$|N_{k,t+1} - N_{k,t}| \le m,$$

so that

$$\left|\frac{N_{k,t+1}}{t+1} - \frac{N_{k,t}}{t}\right| \le \frac{m+1}{t}.$$

Hence,

$$\begin{aligned}\mathrm{Var}\left(\frac{N_{k,t+1}}{t+1}\bigg|\mathcal{G}_t\right) &= \mathbb{E}\left(\left(\frac{N_{k,t+1}}{t+1}\right)^2\bigg|\mathcal{G}_t\right)-\mathbb{E}\left(\left(\frac{N_{k,t+1}}{t+1}\right)\bigg|\mathcal{G}_t\right)^2\\ &\le \left(\frac{m+1}{t}\right)^2.\end{aligned}$$

Therefore, for $k > m$

$$\begin{aligned}&\Bigg|\mathbb{E}\left(\left(\frac{N_{k,t+1}}{t+1}\right)^2|\mathcal{G}_t\right)-\left(\frac{N_{k,t}}{t}\right)^2-\frac{2}{t+1}\left(\frac{m}{c_t}\right)\\ &\times\left(\left(k+\frac{c_t}{m}\right)\left(\left(\frac{k-1}{k+\frac{c_t}{m}}\right)\frac{N_{k-1,t}}{t}\frac{N_{k,t}}{t}-\left(\frac{N_{k,t}}{t}\right)^2\right)\right)\Bigg| < D_k t^{-2},\end{aligned}\tag{4.16}$$

where D_k is a constant depending only on k. If $k = m$, then

$$\begin{aligned}&\Bigg|\mathbb{E}\left(\left(\frac{N_{m,t+1}}{t+1}\right)^2|\mathcal{G}_t\right)-\left(\frac{N_{m,t}}{t}\right)^2-\frac{2}{t+1}\left(\frac{m}{c_t}\right)\\ &\times\left(\left(m+\frac{c_t}{m}\right)\left(\left(\frac{c_t/m}{m+\frac{c_t}{m}}\right)\frac{N_{m,t}}{t}-\left(\frac{N_{m,t}}{t}\right)^2\right)\right)\Bigg| < D_m t^{-2},\end{aligned}\tag{4.17}$$

where D_m is a constant depending only on m.

Consider the case when $m = k$. As

$$\lim_{t\to\infty}\mathbb{E}\left(\frac{N_{m,t}}{t}\right)=\frac{2}{m+2},$$

by taking expectations on both sides of (4.17), we may use Lemma 4.5 with

$$x_t=\mathbb{E}\left(\left(\frac{N_{m,t}}{t}\right)^2\right),$$

$$y_t=\frac{2}{m+2}\mathbb{E}\left(\frac{N_{m,t}}{t}\right)$$

to derive that

$$\lim_{t\to\infty}\mathbb{E}\left(\left(\frac{N_{m,t}}{t}\right)^2\right)=\left(\frac{2}{m+2}\right)^2.$$

Therefore,

$$\begin{aligned}\lim_{t\to\infty}\mathrm{Var}\left(\frac{N_{m,t}}{t}\right) &= \lim_{t\to\infty}\mathbb{E}\left(\left(\frac{N_{m,t}}{t}\right)^2\right)-\mathbb{E}\left(\frac{N_{m,t}}{t}\right)^2\\ &= 0.\end{aligned}$$

For an induction hypothesis, suppose that

$$\lim_{t\to\infty} \mathrm{Var}\left(\frac{N_{k-1,t}}{t}\right) = 0.$$

Then since $\mathrm{Var}\left(\frac{N_{k,t}}{t}\right) \leq 1$, we have that

$$\begin{aligned}\lim_{t\to\infty} &\left|\mathbb{E}\left(\frac{N_{k-1,t}}{t}\right)\mathbb{E}\left(\frac{N_{k,t}}{t}\right) - \mathbb{E}\left(\frac{N_{k-1,t}}{t}\frac{N_{k,t}}{t}\right)\right| \\ &= \lim_{t\to\infty}\left|\mathrm{Cov}\left(\frac{N_{k-1,t}}{t}, \frac{N_{k,t}}{t}\right)\right| = 0\end{aligned}$$

(see Exercise 9). In particular,

$$\mathbb{E}\left(\frac{N_{k-1,t}}{t}\right)\mathbb{E}\left(\frac{N_{k,t}}{t}\right) \sim \mathbb{E}\left(\frac{N_{k-1,t}}{t}\frac{N_{k,t}}{t}\right).$$

Taking expectations on both sides of (4.16), and defining

$$x_t = \mathbb{E}\left(\left(\frac{N_{k,t}}{t}\right)^2\right),$$

$$y_t = \left(\frac{k-1}{k+\frac{c_t}{t}}\right)\mathbb{E}\left(\frac{N_{k-1,t}}{t}\right)\mathbb{E}\left(\frac{N_{k,t}}{t}\right),$$

we have that

$$\begin{aligned}\lim_{t\to\infty} y_t &= \lim_{t\to\infty}\left(\frac{k-1}{k+\frac{c_t}{t}}\right)\mathbb{E}\left(\frac{N_{k-1,t}}{t}\right)\mathbb{E}\left(\frac{N_{k,t}}{t}\right) \\ &= \left(\frac{2m(m+1)}{k(k+1)(k+2)}\right)^2.\end{aligned}$$

By Lemma 4.5

$$\lim_{t\to\infty}\mathbb{E}\left(\left(\frac{N_{k,t}}{t}\right)^2\right) = \left(\frac{2m(m+1)}{k(k+1)(k+2)}\right)^2.$$

By Theorem 4.6,

$$\lim_{t\to\infty}\mathbb{E}\left(\frac{N_{k,t}}{t}\right) = \frac{2m(m+1)}{k(k+1)(k+2)},$$

and so

$$\lim_{t\to\infty}\mathrm{Var}\left(\frac{N_{k,t}}{t}\right) = 0. \qquad \square$$

As in the LCD model from Section 4.2.1, Theorems 4.6 and 4.7 demonstrate that the model $\mathcal{G}(m)$ generates graphs whose degree distribution follows a power law with exponent $\beta = 3$. As with the LCD model, this is a drawback of $\mathcal{G}(m)$, since we would prefer that the power law exponent be controlled by our set of parameters. However, with a modification of (4.1), we can obtain any exponent in the interval $(2, \infty)$. More explicitly, choose a fixed $q \in (-m, \infty)$, and redefine the probability (4.1) to be

$$\frac{q + \deg_{G_t}(w)}{\sum_{u \in V(G_t)} \left(q + \deg_{G_t}(u)\right)}.$$

The analysis in [**138**] of this version of the model (Exercise 7) gives a power law with

$$\frac{N_{k,t}}{t} \sim \left(\frac{\left(2 + \frac{q}{m}\right) \Gamma\left(3 + \frac{q}{m} + m + q\right)}{\Gamma\left(2 + \frac{q}{m} + q + m\right) \Gamma\left(m + q\right)} \right) \frac{\Gamma\left(k + q\right)}{\Gamma\left(3 + \frac{q}{m} + k + q\right)},$$

where $\Gamma(x)$ is the gamma function defined as

$$\Gamma(x) = \int_0^\infty t^{x-1} e^{-t} dt.$$

It follows from the properties of the gamma function that this gives a power law with exponent $\beta = 3 + \frac{q}{m}$ (if $q = 0$, then β is the exponent of Theorem 4.6).

4.2.4. The ACL PA model. Almost at the same time as the LCD model emerged, a set of preferential attachment models different from the LCD and $\mathcal{G}(m)$ models were proposed by Aiello, Chung, and Lu [**5**]. In [**5**], four evolving graph models for W were given, originally inspired by *call graphs* (that is, networks of telephone numbers and the calls between them). Three models produce directed graphs, while one generates undirected graphs. These models—called the ACL models—have some advantages over the LCD model, since the power law exponent β may roam over the interval $(2, \infty)$, dependent on the choice of parameters. Further, a disadvantage of LCD and $\mathcal{G}(m)$ models overcome by the ACL model is that they generate graphs with arboricity m. (See Exercise 5.) The underlying idea of the ACL models is to add edges from new vertices by preferential attachment, but also to add edges *between existing vertices* by preferential attachment.

We describe a simplified undirected version of the ACL model which is presented in [**78**]. The sole parameter of this model $\mathcal{G}_{ACL}(p)$ is a real number $p \in [0, 1]$. The graph G_0 is a single vertex with a loop. Define a *vertex-step* as adding a new vertex v, and adding a single edge uv so that u is chosen from existing vertices by preferential attachment. An *edge-step* consists of adding an edge rs where r and s are both independently chosen from existing vertices by preferential attachment (here, r could equal s). To form the graph G_{t+1}, with probability p take a vertex-step, and with probability $1 - p$ take an edge-step. Note that the number of edges in G_t is $t + 1$, while the number of vertices is a random variable. Using the Chernoff bounds, it can be shown that $|V(G_t)|$ is concentrated around its expected value $1 + pt$ (Exercise 6).

LEMMA 4.8. *In the $\mathcal{G}_{ACL}(p)$ model a.a.s. we have that*

$$\mathbb{P}(||V(G_t)| - \mathbb{E}(|V(G_t)|)| > \lambda) \leq \exp\left(\frac{-\lambda^2}{2pt + \frac{2\lambda}{3}}\right).$$

Via a recurrence relation for the expected value and limit theorem akin to Theorem 4.6 above, the ACL has its expected value of the proportion of vertices of degree k at time t following a power law.

THEOREM 4.9. (1) *In $\mathcal{G}_{ACL}(p)$,*

$$\lim_{t\to\infty} \mathbb{E}\left(\frac{N_{1,t}}{t}\right) = \frac{2p}{4-p},$$

and for all $k > 1$,

$$\lim_{t\to\infty} \mathbb{E}\left(\frac{N_{k,t}}{t}\right) = \frac{2p}{4-p}\left(\frac{\Gamma(k)\Gamma\left(2+\frac{2}{2-p}\right)}{\Gamma\left(k+1+\frac{2}{2-p}\right)}\right).$$

(2) *For k sufficiently large,*

$$\lim_{t\to\infty} \mathbb{E}\left(\frac{N_{k,t}}{t}\right) = O\left(k^{-\left(2+\frac{p}{2-p}\right)}\right).$$

To prove Theorem 4.9, we require the following auxiliary lemma on real sequences, which is Lemma 3.1 from [**78**].

LEMMA 4.10. *If $(a_t : t \in \mathbb{N})$, $(b_t : t \in \mathbb{N})$, and $(c_t : t \in \mathbb{N})$ are real sequences satisfying the relation*

$$a_{t+1} = \left(1 - \frac{b_t}{t}\right) a_t + c_t, \tag{4.18}$$

and $\lim_{t\to\infty} b_t = b > 0$ and $\lim_{t\to\infty} c_t = c$, then $\lim_{t\to\infty} \frac{a_t}{t}$ exists and equals $\frac{c}{1+b}$.

Proof. Fix $\varepsilon > 0$. From (4.18) it follows that

$$\begin{aligned}
&\left|\frac{a_{t+1}}{t+1} - \frac{c}{1+b}\right| \\
&= \left|\frac{\left(1-\frac{b_t}{t}\right)a_t + c_t}{t+1} - \frac{c}{1+b}\right| \\
&= \left|\left(\frac{a_t}{t} - \frac{c}{1+b}\right)\left(1 - \frac{1+b_t}{t+1}\right) + \frac{(1+b)c_t - (1+b_t)c}{(1+b)(1+t)}\right| \\
&\leq \left|\frac{a_t}{t} - \frac{c}{1+b}\right|\left|1 - \frac{1+b_t}{t+1}\right| + \left|\frac{(1+b)c_t - (1+b_t)c}{(1+b)(1+t)}\right|.
\end{aligned} \tag{4.19}$$

By hypothesis, there is a $t_0 \in \mathbb{N}$ such that for $t \geq t_0$

$$|(1+b)c_t - (1+b_t)c| < \varepsilon,$$

$b/2 < b_t$, and $1 + b/2 < t$. Then for $t \geq t_0$ we have by (4.19) that

$$\begin{aligned} \left|\frac{a_{t+1}}{t+1} - \frac{c}{1+b}\right| - \varepsilon &< \left(\left|\frac{a_t}{t} - \frac{c}{1+b}\right| - \varepsilon\right)\left(1 - \frac{1+b/2}{t}\right) \\ &< \prod_{i=1}^{t}\left(1 - \frac{1+b/2}{i}\right), \end{aligned}$$

where the second inequality follows by induction.

As

$$\prod_{i=1}^{t}\left(1 - \frac{1+b/2}{i}\right) \sim \exp\left(-b/2\sum_{i=i}^{t}\frac{1}{i}\right)$$

tends to 0, there is a $t_1 \in \mathbb{N}$ such that for $t \geq t_1$

$$\left|\frac{a_{t+1}}{t+1} - \frac{c}{1+b}\right| < 2\varepsilon.$$

As ε was arbitrary, the lemma is proved. □

Proof of Theorem 4.9. For item (1), we derive a recurrence for $\mathbb{E}(N_{k,t})$. Let $\mathcal{G}_t$ be the graphs $\{G_0, \ldots, G_t\}$. For $k > 1$, $t > 0$, by the definition of the model we have that

$$\begin{aligned} \mathbb{E}(N_{k,t}|\mathcal{G}_{t-1}) &= N_{k,t-1}\left(1 - \frac{pk}{2t} - \frac{(1-p)2k}{2t}\right) \\ &\quad + N_{k-1,t-1}\left(\frac{p(k-1)}{t} + \frac{(1-p)2(k-1)}{2t}\right) \\ &= N_{k,t-1}\left(1 - \frac{(2-p)k}{2t}\right) \\ &\quad + N_{k-1,t-1}\left(\frac{(2-p)(k-1)}{2t}\right). \end{aligned} \tag{4.20}$$

By taking the expectations of both sides of (4.20), we have that

$$\mathbb{E}(N_{k,t}) = \mathbb{E}(N_{k,t-1})\left(1 - \frac{(2-p)k}{2t}\right) + \mathbb{E}(N_{k-1,t-1})\left(\frac{(2-p)(k-1)}{2t}\right).$$

For $k = 1$ and $t > 0$, we have that

$$\mathbb{E}(N_{1,t}|\mathcal{G}_{t-1}) = \mathbb{E}(N_{1,t-1})\left(1 - \frac{2-p}{2t}\right) + p.$$

We proceed by induction to show that $\lim_{t\to\infty}\mathbb{E}\left(\frac{N_{k,t}}{t}\right)$ exists and has limit L_k. For $k = 1$, let $a_t = \mathbb{E}(N_{1,t})$, $b_t = \frac{2-p}{2}$, and $c_t = p$. By Lemma 4.10 we have that

$$L_1 = \lim_{t\to\infty} \mathbb{E}\left(\frac{N_{1,t}}{t}\right) = \frac{2p}{4-p},$$

so this case follows.

Now suppose that $\lim_{t\to\infty} \mathbb{E}\left(\frac{N_{k-1,t}}{t}\right) = L_{k-1}$, for $k \geq 1$. Let $a_t = \mathbb{E}(N_{k,t})$, $b_t = \frac{(2-p)k}{2}$, and $c_t = \mathbb{E}(N_{k-1,t-1})\frac{(2-p)(k-1)}{2t}$. Then by the induction hypothesis,

$$\lim_{t\to\infty} c_t = L_{k-1}\frac{(2-p)(k-1)}{2}.$$

Hence, by Lemma 4.10 we have that

$$\lim_{t\to\infty} \mathbb{E}\left(\frac{N_{k,t}}{t}\right) = L_k = \frac{L_{k-1}(k-1)}{k+\frac{2}{2-p}}. \tag{4.21}$$

The recursion (4.21) allows us to conclude that

$$\lim_{t\to\infty} \mathbb{E}\left(\frac{N_{k,t}}{t}\right) = \frac{2p}{4-p}\left(\frac{\Gamma(k)\Gamma\left(2+\frac{2}{2-p}\right)}{\Gamma\left(k+1+\frac{2}{2-p}\right)}\right),$$

which finishes the proof of (1).

For (2), if L_k is proportional to $k^{-\beta}$ for large k, then

$$\frac{L_k}{L_{k-1}} = \left(1-\frac{1}{k}\right)^{\beta} = 1-\beta\frac{1}{k}+O(1/k^2).$$

The result then follows from (4.21), since

$$\frac{L_k}{L_{k-1}} = \frac{k-1}{k+\frac{2}{2-p}} = 1-\frac{1+\frac{2}{2-p}}{k+\frac{2}{2-p}} = 1-\left(1+\frac{2}{2-p}\right)\frac{1}{k}+O(1/k^2). \qquad \square$$

By exploiting generalized martingale inequalities, a non-trivial concentration result for the ACL model can be derived (see Theorem 3.2 of [**78**] for a proof).

THEOREM 4.11 ([**78**]). *In $G(p)$, a.a.s.*

$$N_{k,t} = L_k t + O(2\sqrt{k^3 t\log(t)}),$$

where $L_1 = \frac{2p}{4-p}$ and

$$L_k = \frac{2p}{4-p}\frac{\Gamma(k)\Gamma\left(2+\frac{2}{2-p}\right)}{\Gamma\left(k+1+\frac{2}{2-p}\right)}$$

for $k \geq 2$.

Hence, a.a.s. $G(p)$ generates graphs possessing power law degree distributions, with power law exponent $\beta = 2 + \frac{p}{2-p}$. Depending on p, the power law exponent β can be any real number in $[2, 3]$.

There are other important preferential attachment models such as the model of Cooper and Frieze [**80**]. Their model is fairly complex, owing to its large number of parameters. Their proof of a power law degree distribution for graphs generated by their model is novel, since it uses martingale techniques along with the Laplace method for the solution of linear difference equations.

4.2.5. The copying model. We saw in Section 4.2.1 the connection between preferential attachment and power law degree distributions. In this section, we consider an evolving graph model that uses a paradigm different from preferential attachment, but nevertheless with high probability generates graphs with power law degree distributions. The motivation for this model is that new pages are formed with only *local knowledge*: they will copy links from an existing vertex, and include a certain number of links chosen from anywhere on the web. The dependence on global knowledge of all vertex degrees is one drawback with PA models.

The *copying model* was introduced in Kleinberg et al. [**141**] and rigorously analyzed in Kumar et al. [**145**]. It exploits local structure to derive a power law degree distribution. It has parameters $p \in (0, 1)$, $d \in \mathbb{N}^+$, and a fixed finite directed graph H with constant out-degree d. Assume that G_t has constant out-degree d. At time $t + 1$, an existing vertex, which we refer to as u_t, is chosen u.a.r. from the set of all existing vertices. The vertex u_t is called the *copying vertex*. For each of the d out-neighbours w of u_t with probability p, add a directed edge (v_{t+1}, z), where z is chosen u.a.r. from $V(G_t)$, and with the remaining probability $1 - p$ add the directed edge (v_{t+1}, w).

For example, suppose z is a new hockey web page. The author—say Sally—of z looks around for an existing hockey web site u_t, and copies some of its links. Sally includes a few other randomly chosen links to round out z, which otherwise would perfectly copy the out-neighbourhood of u_t. See Figure 4.5 for an example.

Kumar et al. use martingales and the Azuma-Hoeffding inequality to prove the following.

THEOREM 4.12 ([**145**]). *If* $k > 0$, *then in the copying model with parameter* p, *a.a.s.*

$$\frac{N^{in}_{k,t}}{t} = \Theta\left(k^{-\frac{2-p}{1-p}}\right),$$

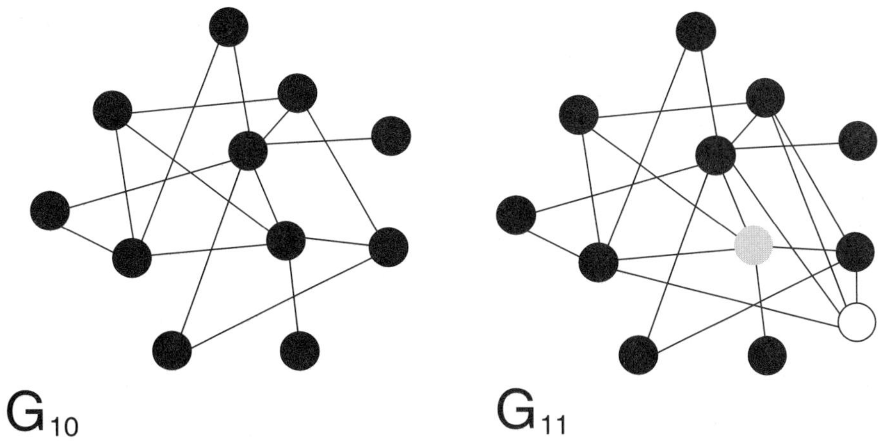

Figure 4.5. A time-step in the copying model. The new white vertex copies from the grey vertex.

where $N_{k,t}^{in}$ is the number of vertices of in-degree k at time t. In particular, the copying model generates directed graphs G_t whose in-degree distribution converges to a power law with exponent

$$\beta = \frac{2-p}{1-p} \in (2, \infty).$$

As described in Chapter 2, the presence of many dense bipartite subgraphs is a desirable property for graphs generated by a web graph model. Kumar et al. [**145**] analyze the model of Aiello et al. [**4**] (which was defined historically before the ACL models) and demonstrate that this model generates graphs which on average contain few bipartite cliques. Two subgraphs of a graph are *distinct* if they have distinct vertex sets. Let $N_{t,i,j}$ denote the expected number of distinct $K_{i,j}$'s which are subgraphs of G_t.

THEOREM 4.13 ([**145**]). *A.a.s. in the copying model with constant out-degree $d > 0$, for $i \leq \log t$,*

$$N_{t,i,d} = \Omega(t \exp(-i)).$$

4.2.6. Growth-deletion models. In all of the models we presented in Sections 4.2.1 and 4.2.5, at each time-step vertices and edges are added, but never deleted. An evolving graph model incorporating in its design both the addition and deletion of vertices and edges may more accurately model the evolution of the web graph. After all, pages and links routinely vanish over time. By what mechanism are they deleted? How can we model such phenomena in W?

One approach to this was adopted by Bollobás and Riordan [**36**], who consider the effect of deleting a set of vertices *after* vertices have been generated in the LCD model. The purpose of this study was to investigate the robustness of graphs generated by the LCD model to random failures, and the vulnerability of these graphs to random attacks. We now describe two recent models, developed independently of each other, that incorporate the addition and deletion of vertices *during* the generation of vertices. Following [**77**], we refer to such models as *growth-deletion models.*

We first describe the growth-deletion model of Chung and Lu [**77**]. They introduce a model $G(p_1, p_2, p_3, p_4, m)$, with parameters m a positive integer, and probabilities $p_1, p_2, p_3, p_4 \in (0, 1)$ satisfying $p_1+p_2+p_3+p_4 = 1$, $p_3 < p_1$, and $p_4 < p_2$; the graph H is a fixed non-empty graph. To form $G_{t+1,}$ we proceed as follows. With probability p_1, add v_{t+1} and m edges from v_{t+1} to existing vertices chosen by preferential attachment. With probability p_2, add m new edges with endpoints to be chosen among existing vertices by preferential attachment. With probability p_3, delete a vertex chosen u.a.r. from $V(G_t)$. With probability p_4, delete m edges chosen u.a.r.

Chung and Lu [**77**] prove the following result by coupling with the offline model $G(\mathbf{d})$ (described in Section 4.2.8 below).

THEOREM 4.14 ([**77**]). (1) *A.a.s. the degree distribution of a graph G_t generated by $G(p_1, p_2, p_3, p_4, m)$ follows a power law distribution with exponent*

$$\beta = 2 + \frac{p_1 + p_3}{p_1 + 2p_2 - p_3 - 2p_4}.$$

(2) *Suppose that $m > \log^{1+\epsilon} t$. Then a.a.s. G_t is connected with*

$$\mathrm{diam}(G_t) = \Theta(\log t).$$

Another recent growth-deletion model developed independently of [**77**] is the one of Cooper, Frieze, and Vera [**81**]. We present a simplified version of the model, disregarding one parameter which controls the deletion of edges. The parameters for this model are a positive integer m and fixed p_1 and p_2 in $(0, 1)$ satisfying $p_2 \le p_1$. The graph G_0 is K_1. With probability $1 - p_1$, delete a vertex of G_{t-1} chosen u.a.r. If G_{t-1} has no vertices, then do nothing. With probability p_2, add m edges from v_{t+1} joined to existing vertices chosen by preferential attachment. The graph is made simple by deleting multiple edges. If there are no edges nor vertices in G_{t-1}, then begin again at time $t = 0$. If there are no edges but some vertices in G_{t-1}, then add v_{t+1} joined to an existing vertex chosen u.a.r. With probability $p_1 - p_2$, add m edges between existing vertices, with endpoints chosen by preferential attachment. The graph is made simple by deleting multiple edges and deleting any loops.

If there are no edges nor vertices in G_{t-1}, then begin again at time $t = 0$. If there are no edges but some vertices in G_{t-1}, then do nothing.

Let

$$\gamma = \frac{2p_1}{3p_1 - 1 - p_2} \text{ and } \rho = \frac{p_2}{p_1}.$$

Cooper, Frieze, and Vera prove the following result on the degree distribution of the model.

THEOREM 4.15 ([**81**]). *Assume that* $p_1 + p_2 > 1$ *and* $3p_1 - p_2 > 1$. *Then a.a.s. there exists a constant* $C = C(m, p_1, p_2)$ *such that for* $k \geq 1$ *and* $1/2 < p_1 \leq 1$,

$$\left| \mathbb{E}\left(\frac{N_{k,t}}{t}\right) - Ck^{-1-\gamma} \right| = O\left(t^{-\rho/8}\right) + O(k^{-2-\gamma}).$$

As noted in [**81**], with a suitable choice of p_1 and p_2, γ may take any value in the interval $(1, \infty)$, and so there is a power law for this model with exponent $\beta = 1+\gamma \in (2, \infty)$. An intriguing problem mentioned in this paper is to rigorously analyze the degree distributions of growth-deletion models where the choice of vertices and edges to delete is not made u.a.r.

4.2.7. Geometric web graph models. In *geometric random graph models*, vertices are identified with points in some geometric space $\mathcal{S}$, and edges are determined via a mixture of probabilistic rules and the position of vertices in $\mathcal{S}$. Such models have been extensively studied; see the book [**173**] for more about them.

While the web graph does not live in physical space, W may be viewed as occupying what might be called a *topic-space*, where pages are closer if they have common topics. For example, two news websites would be closer in topic space than a page on golf and a page on graph theory. One approach to modelling W is to exploit a geometric model. An interesting geometric model was recently proposed by Flaxman, Frieze, and Vera (see [**105**] and also [**107, 108**]). In their model, new vertices may only join to vertices within a certain distance apart. Edges are then chosen by preferential attachment. The graphs they generate occupy the surface of a 3-dimensional sphere S with surface area 1 (hence, with radius $\frac{1}{2\sqrt{\pi}}$). For a point u on S and $r > 0$, we consider the *spherical cap* $B_r(u)$ *of radius* r *centred at* u, defined as

$$\{x \in S : ||x - u|| \leq r\},$$

where $||\cdot||$ is the *Euclidean norm* in $\mathbb{R}^3$. Note that r is fixed. If we let A_r be the area of $B_r(u)$, then check that there is a constant $c_0 > 0$ such that $A_r \sim c_0 r^2$.

The parameters of the model consist of a positive integer m, a positive real number r, and $\alpha \geq 0$ which controls the number of loops. Define G_0 to be K_1. To define G_{t+1}, choose a vertex x_{t+1} u.a.r. from S, and add it to G_t. We first introduce some notation. Define

$$V_t(x_t) = V(G_t) \cap B_r(x_{t+1}),$$

and

$$D_t(x_t) = \sum_{v \in V_t(x_t)} \deg_{G_t}(v).$$

Hence, $V_t(x_t)$ are the vertices of $V(G_t)$ in the cap of radius r centred at x_{t+1}, and $D_t(x_t)$ is the degree sum of vertices in $V_t(x_t)$.

Now add m randomly chosen edges (x_{t+1}, y_i), $1 \leq i \leq m$, from x_{t+1} to existing vertices. Each y_i is chosen independently from $V_t(x_t) \cup \{x_{t+1}\}$ so that for all $i = 1, \ldots, m$, and all $v \in V_t(x_{t+1})$,

$$\mathbb{P}(y_i = v) = \frac{\deg_{G_t}(v)}{\max(D_t(x_{t+1}), \alpha m A_r t)}$$

and

$$\mathbb{P}(y_i = x_{t+1}) = 1 - \frac{D_t(x_{t+1})}{\max(D_t(x_{t+1}), \alpha m A_r t)}.$$

Expressed differently, neighbours of x_{t+1} are chosen via preferential attachment, but only from within $B_r(x_{t+1})$. See Figure 4.6.

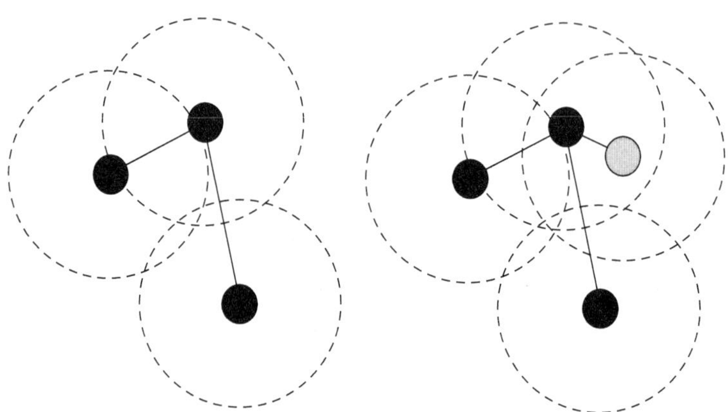

Figure 4.6. A time-step in the evolution of the geometric preferential attachment model when $m = 1$. The grey vertex is the new vertex, while the dotted circles are the caps.

This mix of preferential attachment and geometry leads to a power law, as described in the following theorem.

THEOREM 4.16 ([**105**]). (1) *If* $0 < \beta < \frac{1}{2}$ *and* $\alpha > 2$ *are constants,* $r \sim t^{\beta - \frac{1}{2}} \log t$, *and* m *is sufficiently large, then there exist constants* $c, \gamma, \varepsilon > 0$ *such that for all* $k = k(t) \geq m$,

$$\mathbb{E}\left(\frac{N_{k,t}}{t}\right) = \frac{C_k}{k^{1+\alpha}} + O(t^{1-\gamma})$$

where $C_k = C_k(m, \alpha)$ *tends to a constant* $C(m, \alpha)$ *depending only on* m *and* α *as* k *tends to* ∞.

(2) *The random variable* $N_{k,t}$ *concentrates around its expected value via the following inequality:*

$$\mathbb{P}\left(\left|\frac{N_{k,t}}{t} - \mathbb{E}\left(\frac{N_{k,t}}{t}\right)\right| \geq t^{1-\gamma}\right) \leq \exp(-t^{\varepsilon}).$$

In addition to the power law degree distribution exhibited by Theorem 4.16, the graphs G_t provably low diameter, verifying one part of the small world property.

THEOREM 4.17 ([**105**]). *If* $\alpha \geq 0$, $r \geq t^{-\frac{1}{2}} \log t$, *and* $m \geq K \log t$ *where* K *is sufficiently large, then a.a.s.* G_t *has diameter* $O(\log t / r)$.

The proof of Theorem 4.17 relies on the following technical lemma, whose proof is omitted (see Lemma 6 of [**105**]). Suppose that K_1 is chosen so that $K_1(\log t / A_r) = O(t / \log t)$ and $K_1 = o(K)$, where K is sufficiently large.

LEMMA 4.18 ([**105**]). *If* $u \in S$ *and* $G_t(B)$ *is the subgraph induced by* $V(G_t) \cap B_{r/2}(u)$, *then*

$$\mathbb{P}(\mathrm{diam}(G_t(B)) \geq 2(K_1 + 1) \log t) = O(t^{-3}).$$

Proof of Theorem 4.17. Fix vertices u and v of G_t. Consider a sequence of spherical caps $B_1, \ldots, B_N$ with the following properties:

(1) $N = O(1/r)$.
(2) For $1 \leq i \leq N$, B_i has radius $r/4$.
(3) The centre of B_1 is u, and the centre of B_N is v.
(4) For $1 \leq i \leq N - 1$, the distances between the centres of B_i and B_{i+1} is at most $r/2$.

By items (2) and (4), for $1 \leq i \leq N - 1$, the intersection $B_i \cap B_{i+1}$ has positive area, and so a.a.s. contains a vertex of G_t. To see this, if $p > 0$ is the area of $B_i \cap B_{i+1}$, then the probability that the new vertex x_t at time t is not in $B_i \cap B_{i+1}$ is $1 - p$. Hence, the probability that no vertex is in $B_i \cap B_{i+1}$ is

$$\lim_{t \to \infty} (1 - p)^t = 0.$$

By Lemma 6, a.a.s. $\mathrm{diam}(G_t(B_i)) = O(\log t)$. By items (1) and (3), a.a.s. there is a path connecting u and v in G_t with length at most

$$O(N \log t) = O(\log t/r). \qquad \square$$

A novel feature of the model is that it has *sparse cuts*, which is a property observed in W (see [**28**]). We include the short proof of this fact.

THEOREM 4.19. *If $\alpha \geq 0$ and $r = o(1)$, then a.a.s. $V(G_t)$ can be partitioned into T_1 and T_2 so that $|T_i| \sim \frac{t}{2}$, and there are at most $4\sqrt{\pi}rtm$ edges between T_1 and T_2.*

Proof. Let C be a *great circle* of S (that is, the intersection of S with a plane going through the centre of S), and let T_i be the vertices of G_t below and above C, respectively. Then $|T_1|$ and $|T_2|$ are random variables, both with expected value $\frac{t}{2}$. By the Chernoff bounds, $T_i \sim \frac{t}{2}$.

How many edges cut across C? To estimate this, note that only vertices in the strip X within distance r of C are candidates to cut across C. As $r = o(1)$, X has area less than $3r\sqrt{\pi}$. Let U be the vertices in the strip; by the Chernoff bounds, a.a.s.

$$|U| < 4\sqrt{\pi}rt.$$

The worst case for the number of edges between T_1 and T_2 is that each vertex v of U chooses its m neighbours on the opposite side of the cut, which will yield a.a.s. at most $4\sqrt{\pi}rtm$ edges between T_1 and T_2. $\square$

The *Spatial Preferred Attachment* (*SPA*) geometric web graph model [**6**] was recently proposed as an alternative to the geometric model of [**105**]. In the SPA model, each vertex is placed in space and surrounded by an *influence region*. The area of the influence region is determined by the in-degree of the vertex. Unlike the model of [**105**], in each time-step all regions decrease in area as a function of time; this simulates that fact that the topic-space is growing over time. A new vertex v can only link to an existing vertex u if v falls within the influence region of u. If v falls within the influence region of u, then v will link to u with probability p. Hence, the SPA model is based on the preferential attachment principle, but only implicitly: vertices with high in-degree have a large region of influence, and therefore are more likely to attract new links.

We highlight two features that distinguish the SPA model from previous models. First, a new vertex can choose its links purely based on *local* information. The influence region of a vertex can be seen as the region where a web page is *visible*: only web pages that are close enough (in topic) to fall within the influence region will be aware of the given page, and thus have a possibility to link to it. Further, a new vertex links independently to each vertex visible to it. It follows that the new vertex needs no knowledge of the *invisible* part of the graph (such as in-degree of other vertices, or total number of vertices or links) to determine its neighbourhood. Second, since a new vertex links to each visible vertex independently, the out-degree is not a constant nor chosen according to a pre-determined distribution, but arises naturally from the model.

We now give the formal definition of the SPA model. Let S be the surface of the sphere of area 1 in $\mathbb{R}^3$. The SPA model has parameters $A_1, A_2, p \geq 0$ such that $p \in (0,1]$, $A_1 \leq 1$ and $A_2 \geq 0$. Let $G_t = (V_t, E_t)$, and $V_t \subseteq S$. Let $d^-(v,t)$ be the in-degree of a vertex v in G_t. We define the *influence region* of a vertex v at time $t \geq 1$, written $R(v,t)$, to be the cap around v with area

$$|R(v,t)| = \frac{A_1 d^-(v,t) + A_2}{t},$$

or $R(v,t) = S$ if the right-hand side is greater than 1.

The process begins at $t = 0$, with G_0 being the empty graph, and we let G_1 be just K_1. Time-step t, $t \geq 2$, is defined to be the transition between G_{t-1} and G_t. At the beginning of each time-step t, a new vertex v_t is chosen uniformly at random from S, and added to V_{t-1} to create V_t. Next, independently, for each vertex $u \in V_{t-1}$ such that $v_t \in R(u, t-1)$, a directed edge (v_t, u) is created with probability p. Thus, the probability that a link (v_t, u) is added in time-step t equals $p|R(u, t-1)|$.

Note that the model may be defined on any compact set of measure 1. However, if the set has non-empty boundary, the definition of the influence regions should be adjusted. If higher dimensions are desired, then S could be chosen to be the boundary of a hypersphere in $\mathbb{R}^k$ for some $k > 3$. See Figure 4.7 for a drawing of a graph generated by a simulation of the SPA model (with vertices on the unit square) with 5000 vertices, $p = A_1 = 1$, and $A_2 = 0$.

For an integer $n \geq 0$, define

$$k_f = k_f(t) = \left(\frac{t}{\log^4 t}\right)^{pA_1/(6pA_1+2)}. \tag{4.22}$$

The main result of [**6**] is that with high probability a graph G_t generated by the SPA model has an in-degree distribution that follows a power law in-degree distribution with exponent $1 + \frac{1}{pA_1}$, with concentration up to t^{k_f}. If $pA_1 = 10/11$, then the power law in-degree exponent is 2.1, the same as observed in the web graph (see Chapter 2). Recall that an event holds *with extreme probability* (*w.e.p.*) if it holds with probability at least $1 - \exp(-\Theta(\log^2 t))$ as $t \to \infty$.

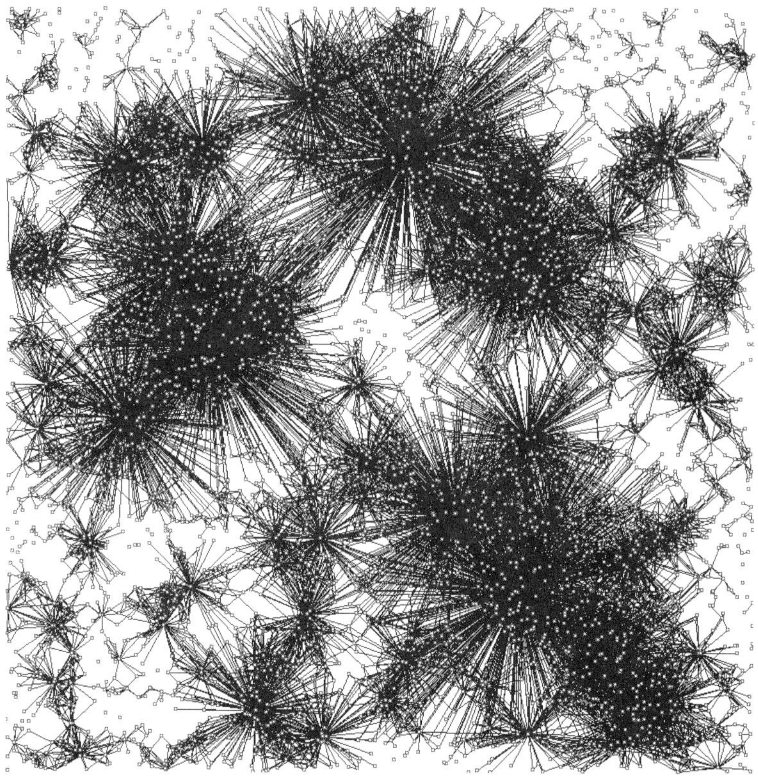

Figure 4.7. A simulation of the SPA model.

THEOREM 4.20 ([**6**]). *Fix $p \in (0,1]$. Then for any $k \geq 0$,*

$$\mathbb{E}(N_{k,t}) = (1+o(1))c_k t,$$

where

$$c_0 = \frac{1}{1+pA_2},$$

and for $1 \leq k \leq n$,

$$c_k = \frac{p^k}{1+pA_2+kpA_1}\prod_{j=0}^{k-1}\frac{jA_1+A_2}{1+pA_2+jpA_1}.$$

For $k = 0, \ldots, k_f$, w.e.p.

$$N_{k,t} \quad = \quad (1+o(1))c_i t.$$

Since it can be shown that $c_k = (1+o(1))ck^{-(1+\frac{1}{pA_1})}$ for some constant c, Theorem 4.20 shows that for large k, the expected proportion $N_{k,t}/t$ follows a power law with exponent $1+\frac{1}{pA_1}$, with concentration for all values of k up to k_f.

4.2.8. Off-line models and quasi-randomness. There are many more on-line models of W that have been rigorously analyzed than we describe here. For example, there is growing interest in rank-based attachment models, where new vertices choose to join to existing vertices on the basis of some measure of prestige (such as age or in-degree). See [**110, 136, 157**]. Some other models of importance that were rigorously analyzed are: Berger et al. [**25**], Bollobás et al. [**33**], Buckley and Osthus [**56**], Aldous [**10**], and Wang [**193**].

While on-line models have dominated our discussion, we should mention one important off-line model of W: the $G(\mathbf{d})$ model designed by Chung and Lu. We take a moment and describe this here. A much more complete description and analysis of this model can be found in [**78**].

The traditional random model $G(n,p)$ may be generalized as follows. Let

$$\mathbf{d} = (d_1, \ldots, d_n)$$

be a sequence of non-negative real numbers. The *configuration model* of Molloy and Reed [**166**] used random matchings to generate random graphs with degree sequence $\mathbf{d}$. We define a model for random graphs with *expected* degree sequence $\mathbf{d}$, written $G(\mathbf{d})$, as follows. Assume that for all $1 \leq i \leq n$, $d_i{}^2 \leq \sum_{k=1}^n d_k$. The edge between v_i and v_j is chosen independently with probability p_{ij} where

$$p_{ij} = \frac{d_i d_j}{\sum_{k=1}^n d_k}.$$

Then $G(n,p)$ may be viewed as a special case of $G(\mathbf{d})$ by taking $\mathbf{d}$ to equal the constant n-sequence $(pn, pn, \ldots, pn)$. In this way, Chung and Lu [**75, 76**] consider $G(\mathbf{d})$ where the expected degree sequence is a power law with fixed exponent β in the interval $(2, \infty)$. They refer to such $G \in G(\mathbf{d})$ as *power law random graphs.* The reader will note that the model $G(\mathbf{d})$ generates off-line graphs, unlike all the other models in this chapter. The motivation for the study of power law random graphs comes in part from the fact that off-line models are easier to work with mathematically than on-line models. A drawback of the model is that the power law is built-in, and does not arise organically from a process such as preferential attachment. Nevertheless, random power law graphs have many properties observed in W. We outline some of elementary properties of $G(\mathbf{d})$ in the exercises at the end of this chapter.

In [**75**], the order of connected components of the graphs in $G(\mathbf{d})$ was investigated. The paper [**76**] proves the following result, which exposes a nice connection between a power law degree distribution and the small world property. We just consider the case $2 < \beta < 3$; for the remaining cases, see Chapter 7 of [**78**].

THEOREM 4.21 ([**75**]). *Suppose that $G \in G(\mathbf{d})$ has n vertices and expected degree sequence* $\mathbf{d}$ *following a power law with exponent $2 < \beta < 3$. Let G have average degree $d > 1$ and maximum degree m satisfying*

$$\frac{\log n}{\log\log n} = o(\log m).$$

For all values of $\beta > 2$, with probability 1 as n tends to ∞, the graph G satisfies

$$\mathrm{diam}(G) = \Theta(\log n),$$

and $L(G)$ (the average distance) is at most

$$(2 + o(1))\frac{\log\log n}{\log(1/(\beta - 2))}.$$

Expected power law degree sequences fall into the more general category of *admissible expected degree sequences* introduced in [**76**]. The results of Theorem 4.21 generalize to $G(\mathbf{d})$ with admissible expected degree sequences; see Theorems 1 and 2 of [**76**], and [**78**]. Further results on $G(\mathbf{d})$ will be presented in Chapter 7.

Theorem 3.4 of Chapter 3 on quasi-random graphs may be generalized to the context of graphs with given degree sequence $\mathbf{d}$; see [**74**]. As $\mathbf{d}$ is any fixed graphical sequence, this notion of quasi-randomness specializes to power law graphs. We will write $\mathbf{d} = (d_x : x \in V(G))$, and make the assumption that $d_x > 0$ for all x. For $X, Y \subseteq V(G)$, define

$$e(X,Y) = |(x,y) : x \in X, y \in Y, xy \in E(G)|.$$

For $X \subseteq V(G)$, define $\mathrm{Vol}(X)$ by

$$\mathrm{Vol}(X) = \sum_{x \in X} d_x.$$

Given a walk P with vertices $x_0, \ldots, x_t$, define

$$w(P) = \prod_{1<i<t} \frac{1}{d_{x_i}}.$$

If P has length 1, then $w(P) = 1$. Given a circuit C with vertices $x_0, \ldots, x_t$, define

$$w(C) = \prod_{1 \le i \le t} \frac{1}{d_{x_i}}.$$

The *weighted adjacency matrix* $M = M(G)$ is the $n \times n$ matrix with rows and columns indexed by $V(G)$, with x, y entry 0 if x and y are not joined, and $\frac{1}{\sqrt{d_x d_y}}$ otherwise. It can be shown (see Chapter 5) that the eigenvalues ρ_i of M, where $1 \le i \le n$, satisfy

$$1 = \rho_1 \ge |\rho_2| \ge \cdots \ge |\rho_n|.$$

For $X, Y \subseteq V(G)$ and $t \geq 1$,

$$e_t(X,Y) = \sum_{P \in P_t(X,Y)} w(P),$$

where $P_t(X, y)$ is the set of all walks of length t between $x \in X$ and $y \in Y$.

For simplicity, we will assume that our graphs satisfy the additional property $U_t(C)$, where C is a large real number and $t \geq 1$, which states that for all $x, y \in V(G)$,

$$e_t(x,y) \leq C \frac{d_x d_y}{\mathrm{Vol}(G)}.$$

We now define quasi-randomness properties for an infinite family $G \in G_{\mathbf{d}}(n)$ of graphs which have degree sequence $\mathbf{d}$. Fix $\epsilon \in (0,1)$.

DISC(ϵ) :

For all $X, Y \subseteq V(G)$,

$$\left| e(X,Y) - \frac{\mathrm{Vol}(X)\mathrm{Vol}(Y)}{\mathrm{Vol}(G)} \right| \leq \epsilon \mathrm{Vol}(G).$$

DISC$_t(\epsilon)$:

For all $X, Y \subseteq V(G)$,

$$\left| e_t(X,Y) - \frac{\mathrm{Vol}(X)\mathrm{Vol}(Y)}{\mathrm{Vol}(G)} \right| \leq \epsilon \mathrm{Vol}(G).$$

EIG(ϵ) :

For all $i \geq 1$,

$$|\rho_i| < \epsilon.$$

TRACE$_{2t}(\epsilon)$:

$$\sum_{i \geq 1} \rho_i^{2t} \leq \epsilon.$$

CIRCUIT$_t(\epsilon)$:

$$\left| \sum_{C_t} w(C_t) - 1 \right| \leq \epsilon,$$

where C_t is a circuit of length t.

For properties A and B, we write $A \Rightarrow B$ if for every $\epsilon > 0$, there is a $\delta > 0$ such that $A(\delta)$ implies $B(\epsilon)$. We say that A and B are *equivalent* if $A \Rightarrow B$ and $B \Rightarrow A$.

THEOREM 4.22 ([**74**]). *If G satisfies $U_{t-1}(C)$ for some $t \geq 2$, then* **DISC**, **DISC**$_t$, **EIG**, **TRACE**$_{2t}$, and **CIRCUIT**$_{2t}$ *are all equivalent.*

Hence, under the assumptions U_t, the above properties are all equivalent; a graph satisfying any of them is called a *quasi-random graph with degree sequence* $\mathbf{d}$.

4.3. Future Challenges in Modelling the Web Graph

What remains to be accomplished in the field of modelling W? Several models have been proposed and rigorously analyzed, and several more models are being studied at the present time. However, many challenges remain. In a real sense, research on W has only just begun. According to [**164**], research on power laws and W is divided into five types: observation, interpretation, modelling, *validation* (checking mathematical properties of the model against observed properties of W), and *control* (using information from the model to enhance W). Research on W is moving now from the first three types to more emphasis on validation and control. As an example of this, it is commonly accepted now that the wide-spread use of search engines is altering the composition of web pages. For example, in [**66**], the authors model the impact of search engines on the structure of W.

In this section, we describe some important challenges left in modelling W, and mention possible directions for future research.

1. Simplification and unification of the analysis of models. Analysis of models for W is often difficult, even for relatively simply stated models. It is much easier to introduce a model than to rigorously analyze it. A worthwhile goal in modelling W is the development of a general theory or "calculus" of web graph models, where a few mild initial conditions in the model would ensure that the proportions $\frac{N_{k,t}}{t}$ concentrate on a power law. Such a calculus could play a unifying role in the study of models of W.

2. Vertex-deletion models. Another important challenge in modelling W lies in analyzing models where vertices and edges are deleted over time by more natural rules than just deleting them u.a.r. Anyone who seriously considers modelling W thinks of this almost immediately, but rigorous analysis of such models has proved to be difficult. In the growth-deletion models in Section 4.2.6, edges and vertices are deleted over time u.a.r. What if vertices (or edges) with low degree have a higher probability of being deleted? Such models are easy to formulate and clearly reflect a more accurate picture of W, but the techniques to analyze these models are elusive as yet. Technical difficulties arise in computing $\mathbb{E}(N_{k,t})$ and proving concentration results. Perhaps a calculus of models from item 1 above could be used to remove some of the mathematical hurdles.

3. Generative mechanisms underlying the evolution of W. Other difficulties arise in finding organic graph processes that generate graphs with properties like those of W (such as power law degree distributions and the small world property). So far, preferential attachment and copying seem to be the only alternatives to rigorously generating on-line power law graphs. But preferential attachment has several drawbacks, the main being the need for global knowledge of the degree sequence. Copying models are usually difficult to analyze owing to their rich dependencies. Are there other undiscovered processes at work in W? The likely answer is "yes", and geometric models may lead to new insights concerning these processes.

4. Power law research. A recent paper of Alderson, Doyle, Li, and Willinger [**9**] demonstrates that power law graphs come in all shapes and sizes, and their work suggests a general mathematical framework for their analysis. They single out the physical internet and biological networks such as PPI graphs as exhibiting complex properties not merely explained by the presence of a power law. For example, in the router-level internet graph, the "core" of the network consists of routers whose interconnection is more dependent on various tradeoffs with ISPs than with graph structure. A graph parameter $s(G)$ is introduced in [**9**], which is defined as

$$s(G) = \sum_{ij \in E(G)} \deg(i)\deg(j).$$

For example, $s(K_n) = \binom{n}{2}(n-1)^2$ and $s(K_{1,n}) = n^2$. The parameter $s(G)$ is large when high-degree vertices are pairwise joined. The authors suggest that "scale-free" networks should not only be power law graphs, but should have large s-value.

An open problem is to design a rigorous graph model which simulates properties of physical networks such as the various levels of the internet graph. For example, any model of the internet should take into account the technological and economic constraints at work (which are more important in this context than in W). An important lesson derived from [**9**] is the importance of carefully analyzing empirical data when designing models of power law networks.

5. Analysis of graph parameters and structural properties within existing models. Many graph parameters have been successfully studied in $G(n,p)$ random graphs. For instance, the chromatic number of random graphs has been extensively researched (see Chapter 3). While the degree sequence and diameter of web graph models have received the most attention, these are not the most historically important graph-theoretic parameters. Little work has been done on graph parameters for models of W, other than the degree sequence and the diameter. A good example of research in this direction is [**82**], which studied the domination number of graphs generated by a PA model. Besides graph parameters, structural properties of graphs generated by the models need to be further studied. Some preliminary work in this direction on web graph models and infinite graphs is described in Chapter 6.

Exercises

(1) Several of the web graph models in this section generate power law graphs with exponent in an interval of real numbers, depending on the choice of parameters. Confirm that the intervals are correct for the given expression. Assume that $p \in (0,1)$.
  (a) $2+\frac{p}{2-p} \in (2,3)$ (for the ACL model).
  (b) $\beta = \frac{2-p}{1-p} \in (2,\infty)$ (for the copying model).
  (c) If $p_1+p_2+p_3+p_4 = 1$, $p_3 < p_1$, and $p_4 < p_2$, then

$$\beta = 2 + \frac{p_1+p_3}{p_1+2p_2-p_3-2p_4} \in (2,\infty)$$

  (for the Chung and Lu growth-deletion model).
  (d) If $p_1 \geq p_2$ and $3p_1 - p_2 > 1$, then $\beta = 1 + \frac{2p_1}{3p_1-1-p_2} \in (2,\infty)$ (for the Cooper, Frieze, and Vera growth-deletion model).

(2) An n-sequence $\mathbf{d} = (d_1,\ldots,d_n)$ with $d_i \leq d_{i+1}$ for all $1 \leq i \leq n-1$ is *graphical* if it is the degree sequence of some graph of order n.
  (a) Prove the Erdős and Gallai characterization of graphical sequences: $\mathbf{d}$ is graphical if and only if $\sum_{i=1}^{n} d_i$ is even, and for all $k \leq n-1$,

$$\sum_{i=1}^{k} d_i \leq k(k-1) + \sum_{i=k+1}^{n} \min\{k, d_i\}.$$

  (b) Prove the Havel and Hakimi characterization of graphic sequences: if $n \geq 3$ and $d_1 \geq 1$, then $\mathbf{d}$ is graphical if and only if

$$(d_2-1,\ldots,d_{d_1+1}-1, d_{d_1+2},\ldots,d_n)$$

  is graphical.

(3) Show that for t a positive integer, there are $\frac{(2t)!}{t!2^t}$-many distinct LCD's on $[2t]$.

(4) Prove Theorem 4.1.

(5) The *arboricity* of a graph is the minimum number of forests into which its edge set can be partitioned. Prove that for $m \geq 1$, both the LCD model with parameter m and the $\mathcal{G}(m)$ model generate graphs with arboricity m.

(6) Prove Lemma 4.8.

(7) For a fixed $q \in (-m, \infty)$, redefine the probability (4.1) in the definition of the $\mathcal{G}(m)$ model as

$$\frac{q + \deg_{G_t}(w)}{\sum_{u \in V(G_t)} \left(q + \deg_{G_t}(u)\right)}.$$

Prove that this new model a.a.s. generates power law graphs, with exponent

$$\beta = \left(\frac{\left(2 + \frac{q}{m}\right) \Gamma\left(3 + \frac{q}{m} + m + q\right)}{\Gamma\left(2 + \frac{q}{m} + q + m\right) \Gamma\left(m + q\right)} \right) \frac{\Gamma\left(k + q\right)}{\Gamma\left(3 + \frac{q}{m} + k + q\right)},$$

where $\Gamma(x)$ is the gamma function.

(8) Derive (but do not solve) the master equation describing $N_{k,t}$ for the copying model described in Section 4.2.5.

(9) (a) For random variables X and Y on a common probability space, define the *correlation coefficient of* X *and* Y by

$$\rho(X, Y) = \frac{\mathrm{Cov}(X, Y)}{\sqrt{\mathrm{Var}(X)\mathrm{Var}(Y)}}.$$

Prove that $|\rho(X, Y)| \leq 1$.

(b) Let (X_t) and (Y_t) be infinite sequences of random variables on a common probability space. If $\lim_{t\to\infty} \mathrm{Var}(X_t) = 0$ and for some real number $M > 0$, $\mathrm{Var}(Y_t) \leq M$ for all t, then

$$\lim_{t\to\infty} \mathrm{Cov}(X_t, Y_t) = 0.$$

(10) We explore some of the properties of the off-line $G(\mathbf{d})$ model from Section 4.2.8. Assume that $\mathbf{d}$ is a sequence of non-negative real numbers, and define $\rho = 1/\left(\sum_{j=1}^{n} d_j\right)$.

(a) If $\mathbf{d} = (d_1, \ldots, d_n)$, then show that the expected degree of the ith vertex is w_i.

(b) For a graph G in $G(\mathbf{d})$, a.a.s. the expected degree of the ith vertex v_i satisfies

$$|\deg(v_i) - d_i| < c\sqrt{d_i},$$

where c is a constant.

(c) For a graph G in $G(\mathbf{d})$, the probability that G is disconnected is

$$e^{-d_i} + O\left((d_i\rho)^2 \sum_{j=1}^{n} d_j^2 \right),$$

and the expected number of isolated vertices is

$$\sum_{i=1}^{n} \left(e^{-d_i} + O\left((d_i\rho)^2 \sum_{j=1}^{n} d_j^2 \right) \right).$$

(d) For a graph G in $G(\mathbf{d})$ and $k > 0$ an integer, find an asymptotic formula for the expected number of vertices of degree k.

(11) [**74**] For the following exercise, use the notation of Section 4.2.8.

(a) Prove that for all $\epsilon > 0$ and all $t \geq 1$, $\mathbf{CIRCUIT}_{2t}(\epsilon)$ is equivalent to $\mathbf{TRACE}_{2t}(\epsilon)$.

(b) Prove that for all $t \geq 1$, and C a large positive real number, if G satisfies $U_t(C)$, then it satisfies $U_{t+1}(C)$.

Chapter 5

Searching the Web

Search well and be wise, nor believe that self-willed pride will ever be better than good counsel.

–Aeschylus

The ultimate search engine would basically understand everything in the world, and it would always give you the right thing. And we're a long, long ways from that.

–Larry Page

5.1. Introduction

The web contains a seemingly infinite ocean of information. Traversing this ocean unaided would be difficult if not impossible. As most of us experience on a daily basis, search engines are an invaluable tool to surf the vast storehouse of information available on-line. Our focus so far has been exclusively on stochastic models for W. However, modelling represents only one part of the current research focus on the web. Searching the web is an intensively studied area of theoretical computer science. Techniques here tend to more algebraic, although graph theory and discrete probability theory appear many times along the way.

It may be surprising at first to learn that modern search engines exploit the graph structure of the web, but all the algorithms for web searching we present heavily use this structure. In fact, the success of the current generation of search engines is in no small part due to the analysis of the underlying graph-theoretic structure of the web, or *link analysis*. This chapter is an introduction to some of the more popular link analysis algorithms, and we will supply the reader with the requisite mathematical background.

We will first give in Section 5.2 an elementary sketch of how search engines work. Linear algebra, spectral graph theory, and Markov chains are the mathematical tools of choice in this context, so in Sections 5.3 and 5.4 we will survey these three areas. We will then discuss in turn several of the most famous graph searching algorithms: PageRank (the most famous ranking algorithm used in the Google search engine), HITS, and SALSA. As with our previous work, we focus on the mathematics involved in web searching rather than the engineering aspects. For additional reading, the reader is directed to the texts [**19, 65, 149**].

5.2. An Overview of Search Engines

How does a search engine work? A detailed answer is beyond the scope of this book. Nevertheless, a high-level picture of a search engine will be useful to set the stage for the web search algorithms we describe later.

A *search engine* may be broadly defined as an information retrieval system for the web, which when queried on some topic returns a list of documents which share some content or relationship to the topic. Following the discussion in [**131**], a web search engine possesses three major components: a crawler, and indexer, and a query engine. The *crawler* is a computer program that autonomously collects, categorizes, and stores large amounts of raw data from the web. The crawler supplies a connection between the search engine and the web itself; the deeper and more frequent the crawl, the better coverage the search engine has. Data is stored and compressed in an *indexer* which extracts information from the data collected from the crawler (such information could be text or images, for example). The indexer builds an *index*, which is the main data structure used by the search engine. The final component is a *query engine* which responds in real-time to queries from users. As part of the query engine, a *ranking algorithm* attempts to rank web pages in order of their relevance to the query. The ranking is often achieved by the assignment of a weight, or weights, to a given page, and basing the ranking on the order of these weights. Of course, many other components are at play in a state-of-the-art search engine like Google, most of which are not known to the public. For a more detailed introduction to the architecture of search engines, the reader is directed to [**19, 149**]. The most well-known ranking algorithm is PageRank, which we describe in detail in Section 5.5. Regardless of the other ranking tools used in the query engine of Google, it is clear that PageRank plays a significant role. According to Google's website: "... while we have dozens of engineers working to improve every aspect of Google on a daily basis, PageRank continues to provide the basis for all of our web search tools."

The first search engine was Archie, which was created in 1990, and was basically a database of web filenames which could be queried by users. The first commercial search engines came into existence around 1994, with the introduction of engines such as Lycos, Infoseek, and AltaVista. These search engines were text-based, in that they ranked pages higher which had a high frequency of the query phrase. Such engines were susceptible to *term spamming*, where words are repeated in a document to increase its ranking. Around the late 1990's, research on link-based analysis became more intensive, which ultimately led to the development of modern search engines such as Google, Yahoo! Search, and MSN Search in the new millennium.

5.3. Adjacency Matrices and the Perron-Frobenius Theorem

We assume that the reader has some background in elementary matrix algebra. Nevertheless, we give some background and notation for the linear algebraic techniques used in this chapter. Readers familiar with these topics can proceed to the next section. A general reference for matrix theory is [**120**].

Scalars will be real or complex numbers, and will be taken as complex unless otherwise stated; the field of complex numbers is denoted by $\mathbb{C}$. The modulus of a complex number z is denoted $|z|$.

Given an $m \times n$ matrix A with m rows and n columns, the (i,j) entry of A is written $A_{i,j}$. The *transpose* of A is A^T formed by interchanging the rows and columns. A matrix is *symmetric* if $A^T = A$. A matrix is *positive* (*non-negative*) if each entry is positive (non-negative). The $n \times n$ *identity matrix* is written I_n, while the $m \times n$ matrix of all 1's is written $J_{m,n}$ (this notation conflicts with the notation $A_{i,j}$ given above, but we only use it for this special matrix). A 0-1 *matrix* has all entries 0 or 1. The *inverse* of a matrix A (if it exists) is written A^{-1}.

Vectors $\mathbf{x} \in \mathbb{C}^n$ are written in **bold**. The ith entry of an $n \times 1$ vector is written as x_i. We take $\mathbf{x}$ to be a column vector, and represent the corresponding row vector by $\mathbf{x}^T$.

A *norm* is a function $\|\cdot\| : \mathbb{C}^n \to \mathbb{R}$ such that for all vectors $\mathbf{x}$ and $\mathbf{y}$, and scalars c,

(1) $\|\mathbf{x}\| \geq 0$, and $\|\mathbf{x}\| = 0$ if and only if $\mathbf{x}$ is the zero vector.

(2) $\|c\mathbf{x}\| = |c| \, \|\mathbf{x}\|$.

(3) $\|\mathbf{x} + \mathbf{y}\| \leq \|\mathbf{x}\| + \|\mathbf{y}\|$.

The 2-*norm* (or *Euclidean norm*) of a vector $\mathbf{x}$ is defined as

$$\|\mathbf{x}\|_2 = \left(\sum_{i=1}^{n} |x_i|^2 \right)^{1/2}.$$

We will sometimes abuse notation and refer to the 2-norm as simply the norm, and write $\|\mathbf{x}\|_2 = \|\mathbf{x}\|$. A vector is *normalized* (or *unit*) if its norm is 1. The 1-*norm* (or *taxicab norm*) of $\mathbf{x}$ is defined as

$$\|\mathbf{x}\|_1 = \sum_{i=1}^{n} |x_i|.$$

Vectors are *orthogonal* if their dot product

$$\mathbf{x} \cdot \mathbf{y} = \sum_{i=1}^{n} x_i y_i$$

is 0. An *orthogonal* set of vectors consists of a set of pairwise orthogonal vectors. An *orthonormal* set is an orthogonal set where the norm of each vector in the set is 1 (that is, they are all *unit vectors*).

Given an $n \times n$ matrix A, an *eigenvalue* λ of A is a complex number satisfying $A\mathbf{x} = \lambda\mathbf{x}$ for some non-zero vector $\mathbf{x} \in \mathbb{C}^n$. The vector $\mathbf{x}$ is called an *eigenvector* corresponding to λ. The subspace of $\mathbb{C}^n$ consisting of all eigenvectors corresponding to λ is named the *eigenspace* E_λ. Eigenvalues are roots of the characteristic equation

$$\det(xI_n - A) = 0.$$

The *algebraic multiplicity* of an eigenvalue is the power of the monomial $x - \lambda$ in the characteristic equation (which splits over $\mathbb{C}$). The *geometric multiplicity* of λ is the dimension of E_λ. It follows that A has at most n distinct eigenvalues, and the sum of all algebraic multiplicities is n.

A *diagonal matrix* is a square matrix where all off-diagonal entries are 0. An $n \times n$ matrix A is *diagonalizable* if there is an $n \times n$ invertible matrix P and a diagonal matrix D such that $A = P^{-1}DP$. A set of vectors $\{\mathbf{v}_1, \ldots, \mathbf{v}_n\}$ is a *basis* for $\mathbb{C}^n$ if and only if each $\mathbf{u} \in \mathbb{C}^n$ may be written uniquely (up to the order of terms) as

$$\mathbf{u} = \sum_{i=1}^{n} c_i \mathbf{v}_i,$$

where the c_i are scalars. An $n \times n$ matrix A is diagonalizable if and only if there is a basis for $\mathbb{C}^n$ consisting of eigenvectors of A.

PageRank and other ranking algorithms utilize the link structure of the web. A common data structure used to represent W is its adjacency matrix. Any graph may be represented as a matrix in the following way. Given a directed graph G of order n with $V(G) = [n]$, the *adjacency matrix* $A(G)$ is the $n \times n$ matrix whose (i, j) entry is 1 if $ij \in E(G)$, and 0 otherwise. For example, the digraph in Figure 5.1 has the following adjacency matrix:

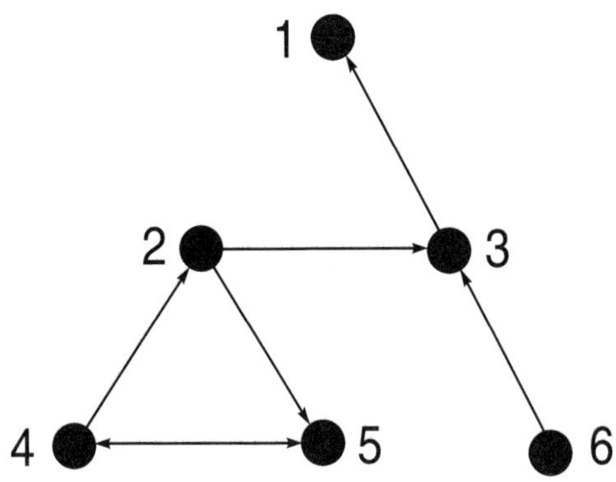

Figure 5.1. A digraph.

$$\begin{pmatrix} 0 & 0 & 0 & 0 & 0 & 0 \\ 0 & 0 & 1 & 0 & 1 & 0 \\ 1 & 0 & 0 & 0 & 0 & 0 \\ 0 & 1 & 0 & 0 & 1 & 0 \\ 0 & 0 & 0 & 1 & 0 & 0 \\ 0 & 0 & 1 & 0 & 0 & 0 \end{pmatrix}.$$

Hence, $A(G)$ is a 0-1 matrix, and is symmetric if G is undirected. If there are no loops in G, then the main diagonal consists of 0's. The adjacency matrix has been intensively studied in graph theory and is related to a number of combinatorial properties of the graph. As a simple example, we have the following lemma whose proof is an exercise.

LEMMA 5.1. (1) *Let G be a directed graph, and let k be a positive integer. Then the number of directed i, j-walks of length k is the (i, j) entry of A^k.*

(2) *Let G be an undirected graph, and let k be a positive integer. Then the number of i, j-walks of length k is the (i, j) entry of A^k.*

A dynamic subfield of graph theory has developed surrounding spectral techniques, studying the eigenvalues and eigenvectors of $A(G)$ or analogous graph-theoretic matrices (such as the Laplacian; see the exercises at the end of this chapter and Chapter 7). A basic but important fact is the following theorem. See [**120**] for a proof.

THEOREM 5.2. *A real, symmetric matrix has all real eigenvalues. In particular, if G is an undirected graph, then $A(G)$ has all real eigenvalues.*

From this fact, for a graph G we may order its eigenvalues $\lambda_1, \ldots, \lambda_n$ by decreasing absolute value:

$$|\lambda_1| \geq \cdots \geq |\lambda_n|.$$

The *spectral radius* of a matrix A is

$$\rho(A) = \max\{|\lambda| : \lambda \text{ is an eigenvalue of } A\}.$$

For an undirected graph, the spectral radius of $A(G)$ is real and equals $|\lambda_1|$. We say that λ_1 is a *dominant* or *first eigenvalue*. If G is directed, then $A(G)$ is not in general symmetric, and so may have complex eigenvalues. If $|\lambda_1| > |\lambda_2|$, then we say that λ_1 is *strictly dominant*.

For an example, let G be K_n. To determine the eigenvalues of K_n, note that $A(G)$ consists of the $n \times n$ matrix of all 1's $J_{n,n}$ minus the $n \times n$ identity matrix I_n. Now x solves $\det(xI_n - J_{n,n}) = 0$ if and only if $x - 1$ solves $\det(xI_n - (J_{n,n} - I_n)) = 0$. As the eigenvalues of $J_{n,n}$ are n with algebraic multiplicity 1 and 0 with algebraic multiplicity $n - 1$, it follows that the eigenvalues of K_n consist of $n - 1$ with algebraic multiplicity 1 and -1 with algebraic multiplicity $n - 1$. Note that $\rho(K_n) = n - 1$.

If G is strongly connected, then $A(G)$ is said to be *irreducible*. In general, a matrix is *irreducible* if for all i and j, there exists some non-negative integer k such that $A^k_{i,j}$ is positive. An irreducible non-negative matrix A is *primitive* if $A^m > 0$ for some $m > 0$. Hence, every primitive matrix is irreducible, but the converse is false in general (for example, consider the adjacency matrix of C_4).

The following theorem is a central one for primitive matrices, and is central to our discussion.

THEOREM 5.3 (Perron-Frobenius [**112, 174**]). *If A is primitive, then the following properties hold.*

(1) *$\rho(A)$ is real and equals an eigenvalue $\lambda_1 > 0$.*
(2) *The eigenvalue λ_1 has algebraic multiplicity* 1 *and geometric multiplicity* 1.
(3) *The eigenvalue λ_1 is the strictly dominant eigenvalue.*
(4) *The eigenvalue λ_1 has a corresponding positive eigenvector.*

By the Perron-Frobenius theorem, if A is primitive, then λ_1 is the unique eigenvalue of A with modulus λ_1 so that $\lambda_1 > |\lambda_2|$. This fact is critical for the power method, which is a numerical approximation method used in the calculation of PageRank. The unique (up to scaling) corresponding eigenvector of λ_1 in the one-dimensional eigenspace E_{λ_1} is the *dominant eigenvector*. We note that if A is irreducible but not primitive, then items

(1), (2), and (4) of Theorem 5.3 hold, but there may be other eigenvalues with the same modulus as λ_1.

We define limits of matrices as *entrywise limits*. More precisely, if (A_t) is a sequence of $m \times n$ matrices, and L is an $m \times n$ matrix, then we write

$$\lim_{t \to \infty} A_t = L$$

if for all $1 \leq i \leq m$, $1 \leq j \leq n$,

$$\lim_{t \to \infty} (A_t)_{i,j} = L_{i,j}.$$

5.4. Markov Chains

In 1907, Markov published a paper [**158**] that began the study of an important new type of random process. In this process, now called a Markov chain in his honour, the outcome of a given experiment can affect the outcome of the next experiment. Our approach to the PageRank algorithm is via Markov chains, which we introduce in this section. Our main example of a Markov chain will be random walks on graphs, which will tie in with our discussion of PageRank in the next section. Hence, our treatment of the subject is by no means exhaustive. A reader with background on Markov chains may safely move on to the next section. For more reading and background on Markov chains, see [**183**].

Fix n a positive integer. A (*discrete-time, time-homogeneous, finite-state*) *Markov chain* M consists of a discrete-time random process $(X_t : t \in \mathbb{N})$, so that each random variable has codomain in the same finite set $S = \{a_0, \dots, a_n\}$, and with the property that for all $n \geq 1$ and $1 \leq t \leq n$,

$$\mathbb{P}(X_t = a_t | X_{t-1} = a_{t-1}, \dots, X_0 = a_0) = \mathbb{P}(X_t = a_t | X_{t-1} = a_{t-1}).$$

In other words, M has *no memory*: a value of X_t only depends on the previous X_{t-1}. We say that the process $(X_t : t \in \mathbb{N})$ is *Markovian*. The set of possible values S of M is called the *state space*, and without loss of generality we will always consider this to be $\{1, \dots, n\}$ where n is an integer. Think of each X_t roughly as a snapshot of the Markov chain at time t.

For all $i, j \in [n]$, define the *transition probability* for M by

$$P_{i,j} = \mathbb{P}(X_t = j | X_{t-1} = i).$$

The (*one-step*) *transition matrix* is the $n \times n$ matrix $P = P(M)$ so that the (i, j) entry of P is $P_{i,j}$. From the definition of a Markov chain, the matrix P is well defined. Each entry of P is non-negative, and the row sums $\sum_{j \in \mathbb{N}} P_{i,j}$ of P are all 1; that is, P is a *stochastic matrix*. An important fact is that any non-negative stochastic matrix corresponds to the transition matrix of some Markov chain. For a given t, define $\mathbf{p}(t)$ to be the probability distribution row vector whose ith entry is the probability that the Markov chain is at state i at time t. Hence, $\mathbf{p}(t-1)P = \mathbf{p}(t)$.

A useful example for us of a Markov chain is a random walk on a connected graph. We imagine a bug moving around a graph G at random from vertex to vertex along edges. At any vertex u, the bug moves to a new vertex by choosing a neighbour of u u.a.r. More precisely, label the vertices with non-negative integers. If at a given time the bug is on vertex i, and i is joined to j, then the transition probability that the bug moves from i to j is $1/\deg(i)$. As the probability that a bug is at any given state only depends on the degree of the current vertex, the process is Markovian.

We may identify any Markov chain with a weighted digraph whose vertices are elements of the state space, and whose directed edges (i,j) correspond to transition probabilities $P_{i,j} > 0$. We call this the *graph representation of the Markov chain*: it will be useful in what follows.

A Markov chain M is *irreducible* if its graph representation is strongly connected. Thus, M is irreducible if and only if its transition matrix is irreducible. The *period* of a vertex u in the digraph G is the greatest common divisor of the lengths of all closed directed walks in G passing through v. If G has no closed directed walks through v, the period of v is said to be 0. If the period of v is 1, then v is said to be *aperiodic*. See Figure 5.2. It is an exercise that all the vertices of a strongly connected digraph have the same period; we call this the *period* of the digraph. A Markov chain is *aperiodic*

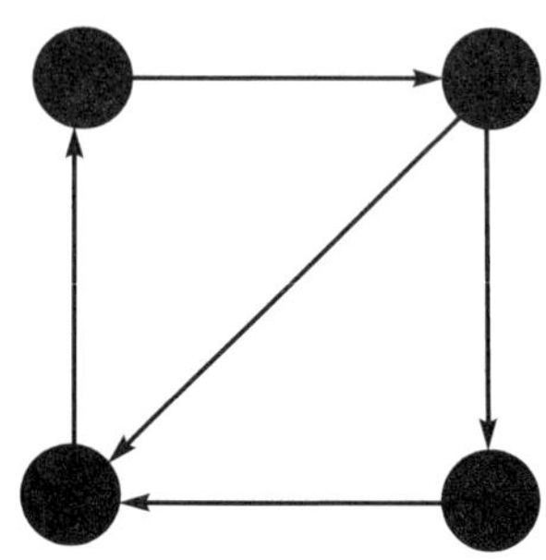

Figure 5.2. An aperiodic digraph.

if all the vertices in its graph representation are aperiodic. An irreducible, aperiodic Markov chain is called *ergodic*. The following fact is useful. See [**183**] for a proof.

LEMMA 5.4. *A Markov chain with primitive transition matrix is ergodic.*

A *stationary distribution* $\mathbf{s}$ of a Markov chain is a *probability distribution* (that is, a vector whose sum of all entries equals 1) with the property that

$\mathbf{s}^T = \mathbf{s}^T P$. We can also express this by saying that $\mathbf{s}$ is an eigenvector of P with eigenvalue 1, or $\mathbf{s}$ is a *fixed point* of P. Stationary distributions always exist if the Markov chain is ergodic. See [**183**] for a proof.

THEOREM 5.5. *An ergodic Markov chain has a unique stationary distribution* $\mathbf{s}$.

According to Theorem 5.5, if we consider an ergodic Markov chain over a long period of time, the initial state becomes increasingly forgotten, and the probability that the state is i approaches the ith component of $\mathbf{s}$. For a proof of Lemma 5.4 and Theorem 5.5, see [**122**]. As we will see, the PageRank vector corresponds to the stationary distribution of a certain Markov chain defined on W.

To illustrate Theorem 5.5, let us consider again our example of random walks on graphs. The reader should verify that a random walk on G is aperiodic if and only if G is not bipartite. A random walk on a finite, connected, non-bipartite graph is therefore an ergodic Markov chain, and so has a stationary distribution $\mathbf{s}$. The following theorem classifies such $\mathbf{s}$.

THEOREM 5.6. *Let G be a finite, connected, non-bipartite graph. A random walk on G converges to a stationary distribution* $\mathbf{s}^T = (s_i)$, *where*

$$s_i = \deg(i)/2|E(G)|.$$

Proof. By the first theorem of graph theory,

$$\sum_{i \in V(G)} s_i = 1.$$

Now, let P be the transition probability matrix of the Markov chain, which has (i, j) entry $1/\deg(i)$. The property $\mathbf{s}^T = \mathbf{s}^T P$ is therefore equivalent to

$$s_i = \sum_{j \in N(i)} \frac{\deg(j)}{2|E(G)|} \frac{1}{\deg(j)} = \frac{\deg(i)}{2|E(G)|}. \qquad \square$$

5.5. PageRank

The most famous and influential ranking algorithm is *PageRank*, discovered by Google's founders Larry Page and Sergey Brin in 1998 [**53**]. Brin and Page were collaborating on their prototype search engine since 1995, and took a leave of absence in 1998 from their computer science doctoral program at Stanford to work on it full time. The rest, including Google's billion dollar initial public offering, is history.

Around the same time, the computer scientist Jon Kleinberg introduced his HITS algorithm [**140**], which while based on link analysis like PageRank, is a completely different approach. We will discuss HITS in the next section.

At its heart, PageRank is simply the stationary distribution of a random walk on a digraph. In other words, the PageRank value of a page is the probability that, at any given moment, a random surfer is visiting this page. How to model this random surfer? According to Brin and Page, most of the time, a surfer will follow links from a page: from a given page i, the surfer will move to one of the out-neighbours of i. A smaller, but positive percentage of time, the surfer will choose a page u.a.r. from W and "teleport" there. We make this rough model of surfer behaviour precise, and show that the random process is a Markovian one which converges to a stationary distribution.

Given $G = W$ and a fixed but arbitrary enumeration of $V(G)$, form the transition matrix P_1 for the random walk on G as described in the last section, but for directed graphs. More explicitly, define

$$P_1(i,j) = \begin{cases} 1/\deg^+(i) & \text{if } (i,j) \in E(G), \\ 0 & \text{else.} \end{cases}$$

A surfer instructed to move to a new page by P_1 has an equally likely chance to move from a vertex to any of its out-neighbours. A problem, unlike the Markov chain for a random walk on either a connected graph or strongly connected digraph, is that the out-degree of a vertex u maybe 0. A surfer only following links then becomes stuck at u!

To overcome this, fix a constant $c \in (0, 1]$. The choice of c by Brin and Page [**53**] was 0.15. Form P_2 by replacing any zero rows of P_1 by the vector $1/nJ_{n,1}$ where $n = |V(G)|$. The vector $1/nJ_{n,1}$ can be replaced with an arbitrary stochastic vector $v > 0$ called the *personalization vector.* Define the *PageRank matrix* (or *Google matrix*) of G by

$$P = P(G) = cP_2 + \frac{(1-c)}{n} J_{n,n},$$

where $J_{n,n}$ is the $n \times n$ matrix of all 1's. For example, consider the graph in Figure 5.2. The PageRank matrix of this graph with vertices 1, 2, 3, and 4 labelled clockwise from the top left is the following:

$$\begin{pmatrix} 17/80 & 29/80 & 17/80 & 17/80 \\ 17/80 & 17/80 & 23/80 & 23/80 \\ 17/80 & 17/80 & 17/80 & 29/80 \\ 29/80 & 17/80 & 17/80 & 17/80 \end{pmatrix}. \tag{5.1}$$

Observe that the PageRank matrix is non-negative. Further, the following lemma establishes that a PageRank matrix is stochastic and hence, is the transition matrix for a Markov chain.

LEMMA 5.7. *For a graph G, the row sums of $P(G)$ are all 1. Hence, $P(G)$ is stochastic.*

Proof. Fix i in $\{1, \ldots, n\}$. We compute the row sum r_i and show it is 1.
Case 1. The out-degree of the vertex i in W is positive.
In this case,

$$\begin{aligned} r_i &= \sum_{1 \leq j \leq n} \left(c(P_2)_{i,j} + \frac{1-c}{n} (J_{n,n})_{i,j} \right) \\ &= c \sum_{1 \leq j \leq n} (P_1)_{i,j} + \frac{1-c}{n} \sum_{1 \leq j \leq n} 1 \\ &= c + (1-c) = 1. \end{aligned}$$

Case 2. The out-degree of i in W is zero.
Now,

$$\begin{aligned} r_i &= \sum_{1 \leq j \leq n} \left(c(P_2)_{i,j} + \frac{1-c}{n} (J_{n,n})_{i,j} \right) \\ &= c \sum_{1 \leq j \leq n} \frac{1}{n} + \frac{1-c}{n} \sum_{1 \leq j \leq n} 1 \\ &= c + (1-c) = 1. \end{aligned}$$

□

By Lemma 5.7, the corresponding Markov chain of $P(W)$ may be viewed as a certain random walk on vertices of W; we call these the *PageRank Markov chain* or *PageRank random walk*, respectively. At a vertex u, with probability c, move to an out-neighbour u.a.r. (if there are no out-neighbours, then move to any vertex in W u.a.r.). With probability $1 - c$, move to any vertex in W chosen u.a.r. The constant c is called the *teleportation constant* (or *damping factor*): it measures how often a surfer teleports to a new page rather than following links.

THEOREM 5.8. *Let* $\mathbf{s}$ *be the dominant eigenvector of* P *with* $\|\mathbf{s}\|_1 = 1$, *corresponding to the eigenvalue* 1. *The PageRank Markov chain with transition matrix* $P = P(G)$ *converges to a stationary distribution equalling* $\mathbf{s}$.

Proof. As each entry of P is positive, P is primitive. Hence, by Lemma 5.4 and Theorem 5.5, the PageRank Markov chain converges to a stationary distribution $\mathbf{s}$. By the Perron-Frobenius Theorem 5.3, $\mathbf{s}$ is the dominant eigenvector with $\|\mathbf{s}\|_1 = 1$. As $\mathbf{s}^T P = \mathbf{s}^T$, we have that the eigenvalue corresponding to $\mathbf{s}$ is 1. □

The vector $\mathbf{s}$ is the *PageRank vector*, and the ith entry is the PageRank of the ith page of G. Although PageRank is used by Google to generate its rankings, it is interesting to note that you must download the Google toolbar to view the PageRank score of a given page in W.

How to compute the PageRank vector $\mathbf{s}$? Since

$$\mathbf{s}^T P = \mathbf{s}^T, \tag{5.2}$$

we must solve the homogeneous linear system

$$(I_n - P)\mathbf{s}^T = 0, \tag{5.3}$$

making sure to normalize so that $\|\mathbf{s}\|_\infty = 1$. Row echelon methods for solving (5.3) are known to anyone who has taken an elementary linear algebra course.

Despite the seeming mathematical simplicity of (5.3), huge computational issues arise when we recall that W is a massive graph, whose order n is about 54 billion. As P has all non-zero entries, the computations for solving (5.3) by standard techniques quickly prove intractable. In fact, PageRank has sometimes been referred to as the world's largest matrix computation.

The good news for solving (5.3) is that a simple technique called the *power method* allows for a quick and effective approximation to the PageRank vector $\mathbf{s}$. The algorithm initializes with a vector $\mathbf{z}_0$, which may be an approximation to the dominant eigenvector. The method works since P is a positive (and hence, primitive) stochastic matrix. Define the sequence $(\mathbf{z}_t : t \in \mathbb{N})$ of vectors by the following recursion. Choose $\mathbf{z}_0$ to be the stochastic vector whose every entry is $1/n$, where $n = |V(G)|$. Assuming $\mathbf{z}_t$ is defined and is stochastic, define

$$\begin{aligned} \mathbf{z}_{t+1}^T &= \mathbf{z}_t^T P \\ &= \left(\mathbf{z}_0^T\right) P^t. \end{aligned}$$

Note that $\mathbf{z}_{t+1}$ is stochastic, since it is the product of two stochastic matrices. We will show that $\lim_{t\to\infty} \mathbf{z}_{t+1}$ is the dominant eigenvector. For the graph in Figure 5.2 with PageRank matrix (5.1), the Pagerank vector after 20 iterations of the power method is approximately

$$\begin{pmatrix} 0.282 \\ 0.277 \\ 0.154 \\ 0.287 \end{pmatrix}.$$

While the power method is slow to converge in general (especially for dense matrices), it is useful for PageRank, as described in [**149**], since P_1 is sparse. To be more precise, let $\mathbf{u}$ be the row-vector defined by

$$u_i = \begin{cases} 0 & \text{if } \deg^+(i) > 0, \\ 1 & \text{else.} \end{cases}$$

It follows that

$$P = cP_1 + (c\mathbf{u} + (1-c)J_{n,1})\frac{1}{n}J_{1,n},$$

and so

$$\begin{aligned} \mathbf{z}_{t+1}^T &= \mathbf{z}_t^T P \\ &= c\mathbf{z}_t^T P_1 + (c\mathbf{z}_t^T \mathbf{u} + (1-c)\mathbf{z}_t^T J_{n,1})\frac{1}{n} J_{1,n}. \end{aligned} \tag{5.4}$$

As the matrix P_1 is sparse (since most web pages have few out-links relative to the massive order of W), the matrix product $c\mathbf{z}_t^T P_1$ in (5.4) is easier to compute. Brin and Page [**53**] reported good results by running between 50 to 100 iterations of the power method. Other numerical methods, such as the Gauss-Seidel and Jacobi method, have been used to compute the PageRank vector. For further discussion, see Section 9.4 of [**149**].

To explain the theory behind the power method, we first rely on the following lemma proved in [**55**].

LEMMA 5.9. *Let M be a positive, stochastic $n \times n$ matrix, and let V be the subspace of $\mathbb{C}^n$ consisting of the set of vectors $\mathbf{z}$ such that*

$$\sum_{i=1}^{n} z_i = 0.$$

Then for all $\mathbf{z} \in V$, the following properties hold.

(1) *$M\mathbf{z} \in V$.*
(2) *$\|M\mathbf{z}\|_1 \leq d\,\|\mathbf{z}\|_1$, for some constant $d \in (0,1)$.*

We now prove a result making the power method rigorous.

THEOREM 5.10. *Let M be a positive, stochastic $n \times n$ matrix, and let $\mathbf{s}$ be the dominant eigenvector of M with $\|\mathbf{s}\|_1 = 1$, corresponding to the eigenvalue 1. Then*

$$\mathbf{s} = \lim_{t\to\infty} M^t \mathbf{z}_0,$$

where $\mathbf{z}_0$ is any positive stochastic vector in $\mathbb{R}^n$.

Proof. We may write $\mathbf{z}_0 = \mathbf{s} + v$, where $v \in V$. Then

$$\begin{aligned} M^t\mathbf{z}_0 &= M^t\mathbf{s} + M^t v \\ &= \mathbf{s} + M^t v. \end{aligned}$$

Hence,

$$\begin{aligned} \left\|M^t\mathbf{z}_0 - \mathbf{s}\right\|_1 &= \left\|M^t\mathbf{v}\right\|_1 \\ &\leq d^t\,\|\mathbf{v}\|_1 \end{aligned}$$

where the last inequality follows from Lemma 5.9 (2) (and $d \in (0,1)$ as in the lemma). Hence,

$$\lim_{t\to\infty} \left\|M^t\mathbf{z}_0 - \mathbf{s}\right\|_1 = 0$$

and so $\mathbf{s} = \lim_{t\to\infty} M^t\mathbf{z}_0$ as desired. □

Experiments conducted in [**192**] reported correlation between PageRank and in-degree in W. Since in-degree is far easier to compute than PageRank, this could open up a vulnerability to *spamdexing* (that is, deliberate spamming attempts to influence Google's ranking of a page; see the final section of this chapter). PageRank is not the same as in-degree for many graphs, as any number of examples demonstrate (see the exercises). However, recent research in [**154**] shows that in digraphs where the in-degree follows a power law, the PageRank distribution of W follows a power law with a similar exponent to the in-degree distribution [**23**].

5.6. HITS

The *hyperlink-induced topic search* (*HITS*) algorithm was introduced by Jon Kleinberg in the late 1990's [**140**], and was one of the first link analysis algorithms for web search. The main idea of the algorithm (which was inspired by certain results in bibliometrics) is that topic can be divided into authorities and hubs. *Authorities* are pages with good coverage of the topic, while *hubs* are pages with links to many useful pages on the topic. HITS identifies good authorities and hubs for a topic, assigning an authority and a hub weight to a page. These weights are defined recursively: a higher authority weight occurs if the page is pointed to by pages with high hub weights, while a higher hub weight results if the page is pointed to many pages with high authority weight. While PageRank is better known, an extension of HITS is used in the search engine Ask.com (see [**149**]).

There are two main components of HITS: a sampling step, and a weight-propagation step. Given a query q, in the *sampling step*, a focused collection of several thousand web pages is generated, which are likely to be relevant to q. We call the subgraph of W induced by these set of pages S, and name it the *base digraph. We will assume that the underlying graph of S (formed by forgetting the orientation of directed edges) is connected.* For example, if we fix k a positive integer, S may be defined as the set of pages at distance at most k from a set of target pages on the topic. In the weight-propagation step, the authority- and hub-values are computed for pages in the base digraph. We direct the interested reader to [**140**] for further details on the generation of S. We will instead focus on the *weight-propagation step.*

Every page p in S will be assigned non-negative real numbers which are the authority and hub weights of p. These weights are calculated by computing vectors $\mathbf{p}(\mathbf{a})$ and $\mathbf{p}(\mathbf{h})$, respectively, which are defined as limit vectors of a certain recursive sequence of vectors. Assuming without loss of generality that $V(S) = [n]$, the authority and hub weight of i can be simply read off as the ith entries in $\mathbf{p}(\mathbf{a})$ and $\mathbf{p}(\mathbf{h})$, respectively.

Now define $\mathbf{p}(\mathbf{a})_0 = \mathbf{p}(\mathbf{h})_0 = J_{1,n}$; this is sometimes called the *seed vector*. For $t \geq 0$, assuming that $\mathbf{p}(\mathbf{a})_t$ and $\mathbf{p}(\mathbf{h})_t$ are defined, define

$$\mathbf{p}(\mathbf{a})'_{t+1} = A(S)^T \mathbf{p}(\mathbf{h})_t.$$

Define

$$\mathbf{p}(\mathbf{h})'_{t+1} = A(S)\mathbf{p}(\mathbf{a})'_{t+1}.$$

Now set

$$\mathbf{p}(\mathbf{a})_{t+1} = \frac{\mathbf{p}(\mathbf{a})'_{t+1}}{\|\mathbf{p}(\mathbf{a})'_{t+1}\|}, \qquad \mathbf{p}(\mathbf{h})_{t+1} = \frac{\mathbf{p}(\mathbf{h})'_{t+1}}{\|\mathbf{p}(\mathbf{h})'_{t+1}\|}.$$

Kleinberg [**140**] proved that under certain assumptions, the sequences

$$(\mathbf{p}(\mathbf{a})_t : t \in \mathbb{N}), \quad (\mathbf{p}(\mathbf{h})_t : t \in \mathbb{N})$$

converge to the dominant eigenvectors of certain matrices, and these limits are the vectors $\mathbf{p}(\mathbf{a})$ and $\mathbf{p}(\mathbf{h})$, respectively.

THEOREM 5.11. *Let $A = A(S)$ be the adjacency matrix of S, and fix $p \in S$. Assuming that $A^T A$ and AA^T are primitive, the following properties hold.*

(1) *The limit $\mathbf{p}(\mathbf{a})$ of the sequence $(\mathbf{p}(\mathbf{a})_t : t \in \mathbb{N})$ is the normalized dominant eigenvector of $A^T A$.*
(2) *The limit $\mathbf{p}(\mathbf{h})$ of the sequence $(\mathbf{p}(\mathbf{h})_t : t \in \mathbb{N})$ is the normalized dominant eigenvector of AA^T.*

Before we prove Theorem 5.11, we need the following version of the power method, where the matrix is diagonalizable.

THEOREM 5.12. *Let M be a non-negative, primitive, diagonalizable matrix, and let $\mathbf{v}_1$ be the dominant eigenvector of M with $\|\mathbf{v}_1\| = 1$. Let $\mathbf{z}_0$ be any vector not orthogonal to $\mathbf{v}_1$. If $\mathbf{z}_t$ is defined, then define*

$$\mathbf{z}_{t+1} = \frac{M\mathbf{z}_t}{\|M\mathbf{z}_t\|}.$$

Then

$$\lim_{t\to\infty} \mathbf{z}_t = \mathbf{v}_1.$$

Proof. By hypothesis, $\mathbb{C}^n$ has a basis of eigenvectors $\{\mathbf{v}_1, \ldots, \mathbf{v}_n\}$, where λ_i is the eigenvalue of v_i, $\lambda_1 = |\lambda_1| > |\lambda_i|$ for all $2 \leq i \leq n$. Hence, we may write

$$\mathbf{z}_0 = \sum_{i=1}^{n} c_i \mathbf{v}_i,$$

where the c_i are scalars, and $c_1 > 0$ as $\mathbf{z}_0$ is not orthogonal to $\mathbf{v}_1$. Observe that

$$\mathbf{z}_t = \frac{M^t \mathbf{z}_0}{\|M^t \mathbf{z}_0\|}.$$

Now,

$$\begin{aligned} M^t \mathbf{z}_0 &= \sum_{i=1}^{n} c_i M^t \mathbf{v}_i \\ &= \sum_{i=1}^{n} c_i \lambda_i^t \mathbf{v}_i \\ &= c_1 \lambda_1^t \left(\mathbf{v}_1 + \sum_{i=2}^{n} \frac{c_i}{c_1} \left(\frac{\lambda_i}{\lambda_1} \right)^t \mathbf{v}_i \right). \end{aligned}$$

As $\left|\frac{\lambda_i}{\lambda_1}\right| < 1$ for all $2 \leq i \leq n$, the sequence $(M^t \mathbf{z}_0 : t \in \mathbb{N})$ converges to a multiple of $\mathbf{v}_1$. As $\mathbf{z}_t$ is the normalized vector of $M^t \mathbf{z}_0$, the proof follows. □

Now we prove the main convergence result for HITS.

Proof of Theorem 5.11. The matrices $A^T A$ and AA^T are real and symmetric, and so are diagonalizable.

Observe that by the definitions, $\mathbf{p}(\mathbf{a})_t$ is the unique unit vector parallel to

$$(A^T A)^{t-1} A^T J_{1,n}$$

and $\mathbf{p}(\mathbf{h})_t$ is the unique unit vector parallel to

$$(AA^T)^t J_{1,n}.$$

We prove item (1), since the proof of item (2) is similar and so is omitted. As $M = A^T A$ is non-negative and primitive, by the Perron-Frobenius theorem, there is a dominant eigenvector $\mathbf{v}_1$ of $A^T A$ with $\|\mathbf{v}_1\| = 1$ (hence, $\mathbf{v}_1$ is non-zero). Let $\mathbf{z}_0 = A^T J_{1,n}$. As $\mathbf{z}_0$ is non-negative with some positive entries, and $\mathbf{v}_1$ is not the zero vector, we have that $\mathbf{v}_1$ and z_0 are not orthogonal. By Theorem 5.12,

$$\begin{aligned} \lim_{t\to\infty} \mathbf{p}(\mathbf{a})_t &= \lim_{t\to\infty} \frac{(A^T A)^{t-1}(A^T J_{1,n})}{\|(A^T A)^{t-1}(A^T J_{1,n})\|} \\ &= \lim_{t\to\infty} \frac{M^{t-1}\mathbf{z}_0}{\|M^{t-1}\mathbf{z}_0\|} \\ &= \mathbf{v}_1. \end{aligned}$$

□

The primitivity hypothesis in Theorem 5.12 is important, since otherwise the resulting limit vectors $\mathbf{p}(\mathbf{a})$ and $\mathbf{p}(\mathbf{h})$ may not be unique. In particular, if the geometric multiplicity of the dominant eigenvalue is more than 1, then $\mathbf{p}(\mathbf{a})$ and $\mathbf{p}(\mathbf{h})$ will depend on the seed vector; see the exercises and Section 5.8. As with PageRank, authority and hub weights may be approximated using the power method.

Let us compute the hub weights of vertices for the digraph in Figure 5.3.

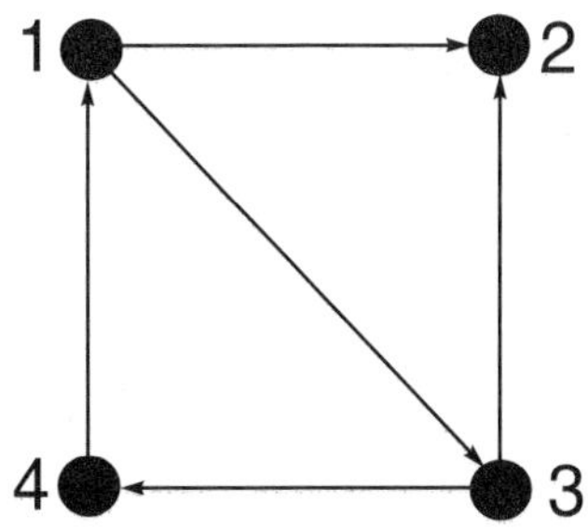

Figure 5.3. The digraph H.

In this case,

$$AA^T = \begin{pmatrix} 2 & 0 & 1 & 0 \\ 0 & 0 & 0 & 0 \\ 1 & 0 & 2 & 0 \\ 0 & 0 & 0 & 1 \end{pmatrix},$$

and the dominant eigenvalue is 3, with normalized eigenvector

$$\mathbf{p}(\mathbf{h}) = \begin{pmatrix} \frac{1}{\sqrt{2}} \\ 0 \\ \frac{1}{\sqrt{2}} \\ 0 \end{pmatrix}.$$

The hub weight for the vertex i may be read off from the ith entry of $\mathbf{p}(\mathbf{h})$.

Despite some similarities, PageRank and HITS nevertheless have some major differences. For example, unlike PageRank, HITS is *query dependent*: it first generates a set S of pages likely to be relevant to the query, then returns two real numbers for a given page in the base digraph, while PageRank returns a real number for each page in W.

5.7. SALSA

Several other ranking algorithms distinct from HITS and PageRank have been developed; see [**65, 149**] for surveys. We focus on just one other ranking algorithm, SALSA, which combines some features of both HITS and PageRank.

Stochastic Approach for Link Structure Analysis (SALSA) was proposed by Lempel and Moran in [**152**], in an attempt to avoid certain problems as reported with HITS (see [**27**]). One such problem is *topic drift*, where some vertices in the base digraph are not related to the query, and documents with the highest authority and hub weights relate to different topics. Like PageRank, SALSA uses random walks, but like HITS, it computes scores on a proper connected induced subgraph S of W.

Suppose we are given a base digraph S (chosen in an analogous way as in HITS). Define a bipartite (undirected) graph $G = G(S) = (V_a \cup V_h, E)$, with colours

$$V_a = \{x_a : x \in V(S), \deg^-(x) > 0\},$$

$$V_h = \{y_h : y \in V(S), \deg^+(x) > 0\},$$

and

$$E = \{y_h x_a : (y, x) \in E(S)\}.$$

We say that V_a is the set of *authority vertices*, while V_h is the set of *hub vertices*. Consider the digraph H in Figure 5.3. Figure 5.4 represents the graph $G(H)$.

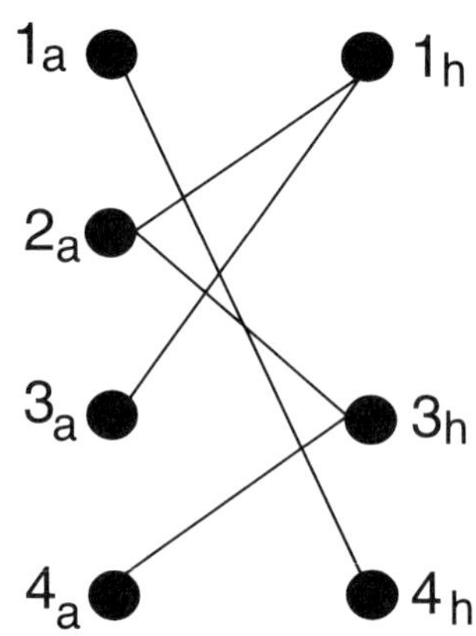

Figure 5.4. The graph $G(H)$.

On the graph G, perform two random walks. Each walk will only visit vertices from one of the two colours by traversing paths consisting of two edges of G in each step. Each walk is therefore restricted to one colour of G, and the two walks start at opposite sides.

More precisely, define an *authority* Markov chain, and a *hub* Markov chain by specifying their respective transition matrices. Following the notation of [**152**], the *hub matrix* H is defined by

$$H_{i,j} = \sum_{\{k : i_h k_a, j_h k_a \in E(G)\}} \frac{1}{\deg(i_h)} \frac{1}{\deg(k_a)},$$

while the *authority matrix* $\overline{A}$ is defined by

$$\overline{A}_{i,j} = \sum_{\{k: k_h i_a, k_h j_a \in E(G)\}} \frac{1}{\deg(i_a)} \frac{1}{\deg(k_h)}.$$

Both matrices are stochastic (Exercise 16), and so represent Markov chains: the authority and hub Markov chains for SALSA.

Let $A = A(S)$, and recall that r_i is the ith row sum of A, while c_i is the ith column sum of A. Let A_r be defined from A by dividing each non-zero (i, j) entry by r_i, and let A_c be defined from A by dividing each non-zero (i, j) entry by c_i. It follows that $\overline{A} = A_c^T A_r$ and $H = A_r A_c^T$ (see Exercise 16). Lempel and Moran proved the following theorem in a manner similar to the corresponding proofs for PageRank and HITS.

THEOREM 5.13 ([**152**]). *Assuming that G is connected, both $\overline{A}$ and H are stochastic and primitive matrices. In particular, the authority and hub Markov chains both converge to stationary distributions, which are the $\|\cdot\|_1$-normalized dominant eigenvector of $\overline{A}$ and H, respectively.*

Lempel and Moran [**152**] constructed graphs with good authorities, where SALSA gives higher authority weights than HITS. A weakness of SALSA is that, like HITS, it is query-dependent.

5.8. Further Analysis of Web Ranking Algorithms

While present day search engines which use link analysis are far more effective than earlier text-based search engines, several challenges remain in ranking algorithms and in the design of search engines. Henzinger [**131**] outlines several open problems faced by search engines. Some of the main ones are the following.

(1) Uniform sampling of web pages.
(2) Modeling the web graph.
(3) Finding duplicate hosts.
(4) Finding top gainers and losers in data streams.
(5) Finding large dense bipartite graphs.
(6) Understanding how eigenvectors partition the web.

In Chapters 2 and 4 we discussed items (2) and (5). As items (1), (3), and (4) correspond to issues not directly related to ranking, we omit a discussion of these. Item (6) is related to finding so-called balanced decompositions of the web as a directed graph. Such decompositions could lead to an interesting clustering of the web graph or for a topic-specific subgraph.

Spamdexing is the intentional attempt to improve the ranking of a page on a search engine. One kind of spamdexing that influences rankings is *link spam*. For instance, as PageRank is sensitive to in-degree, a simple way to spamdex PageRank is to increase the in-degree of your page. A *link farm* is a dense subgraph of W designed to do just that.

Another issue research direction is analysis and comparison of ranking algorithms. If the matrices associated with the HITS are not primitive, then the eigenvectors need not be unique [**103**]. This difficulty does not arise with PageRank, since the PageRank matrix is always primitive. Following [**103**], a ranking algorithm that assigns hub and authority weights (like HITS or SALSA) is *badly behaved* on a graph G if the final result depends on the initial vectors or if some vertex is given authority (hub) weight zero even though it has in-degree (out-degree) greater than zero.

THEOREM 5.14 ([**103**]). *The HITS algorithm is badly behaved on G, with $A = A(G)$, if and only if at least one of the following holds:*

(1) *AA^T has a repeated largest eigenvalue.*

(2) *The dominant eigenvector of AA^T (A^TA) has zero entries for vertices whose out-degree (in-degree) is non-zero.*

Given a directed graph G on the vertex set $[n]$, define the (undirected) *HITS authority graph* G' with vertex set those vertices of G that have positive in-degree. Let $xy \in E(G')$ if there exists a vertex z of G such that $(z, x), (z, y) \in E(G)$.

THEOREM 5.15 ([**103**]). (1) *HITS is badly behaved on G if and only if G' is disconnected.*

(2) *SALSA is badly behaved on G if and only if G' is disconnected.*

Another approach to analyzing ranking algorithms is stability. *Stability* measures the effect on rankings if the graph is *perturbed*: edges are added or deleted. [**168**] discovered a bound on the number of edges that can be altered associated with one vertex without affecting the value of HITS. Define $\delta = \lambda_1 - \lambda_2$, the difference between the first and second cigenvalue of A^TA (the so-called *eigengap*), where A is the adjacency graph of the base digraph S. Suppose that S' results from adding or deleting some edges from S. Let $\mathbf{u}$ be the dominant eigenvector of A^TA associated with S, while $\mathbf{u}'$ is analogous vector for S'.

THEOREM 5.16 ([**168**]). *Fix $x \in V(S)$, and let $\varepsilon > 0$ be a real number. Let d be the maximum out-degree of G. Suppose that G' is the graph formed* *by changing at most k edges incident with x, where*

$$k \le \left(\sqrt{d + \frac{\varepsilon\delta}{4 + \sqrt{2}\varepsilon}} - \sqrt{d} \right)^2 .$$

Then

$$\left\| \mathbf{u} - \mathbf{u}' \right\| < \varepsilon.$$

If the eigengap δ is large, then Theorem 5.16 tells us that HITS is insensitive to minor perturbations in the base digraph. For PageRank, [**168**] proved the following result, which applies to perturbations on a set of vertices.

THEOREM 5.17 ([**168**]). *Suppose that G is perturbed to a graph G', and the set of vertices affected by the perturbation is V'. If $\mathbf{s}$ and $\mathbf{s}'$ are the PageRank vectors for G and G', respectively, and $c \in (0, 1]$ is the teleportation constant, then*

$$\left\| \mathbf{s} - \mathbf{s}' \right\|_1 \le \frac{2}{c} \sum_{i \in V'} s_i .$$

Note that if a set of pages in V' have low PageRank, then by Theorem 5.17 the change in PageRank will be small.

Exercises

(1) Prove Lemma 5.1.

(2) Prove that a matrix A is primitive if and only if A is irreducible and $A_{ii} > 0$ for some i.

(3) Find the eigenvectors and eigenvalues of the adjacency matrices for the following graphs.
   (a) A path P_n.
   (b) A cycle C_n.
   (c) A wheel W_n.
   (d) The Petersen graph.

(4) Prove that if G is bipartite with an eigenvalue λ which has algebraic multiplicity m, then $-\lambda$ is also an eigenvalue of G with algebraic multiplicity m.

(5) Show that diam(G) is less than the number of distinct eigenvalues of G.

(6) Suppose that G is a graph with vertices $[n]$. The *Laplacian matrix* L of G is defined by

$$L_{i,j} = \begin{cases} \deg(i) & \text{if } i = j, \\ -1 & \text{if } ij \in E(G), \\ 0 & \text{else.} \end{cases}$$

(a) Find the eigenvalues of the Laplacians of the following graphs: K_n, C_n, and the hypercube Q_n.
(b) Prove that the least eigenvalue of L is 0. Since L is real and symmetric, this proves that we may order the eigenvalues of L as

$$\lambda_1 \geq \cdots \geq \lambda_n = 0.$$

(c) Prove that $\lambda_1 = 2$ if and only a connected component of G is bipartite and non-trivial.

(7) (a) Show that the set of stochastic $n \times n$ matrices forms a monoid under matrix multiplication.
(b) Prove that if A is a stochastic matrix, then $\rho(A) \leq 1$.

(8) (a) Prove that any two vertices in a strongly connected digraph have the same period.
(b) Show that a random walk on a graph G is aperiodic if and only if G is not bipartite.
(c) Give an example of bipartite graph where the random walk does not converge to a stationary distribution.

(9) (a) Let $E(x, y)$ be the expected number of steps needed to reach a vertex x from y in a random walk on a graph G. Prove that for a connected non-bipartite graph G,

$$E(x, x) = \frac{2|E(G)|}{\deg(x)}.$$

(b) The *cover time* C_G of a graph G is the maximum over all vertices x of the expected time to visit all the vertices in G by a random walk starting from x. Show that

$$C_G \leq 4|V(G)||E(G)|.$$

(10) A Markov chain M with an $n \times n$ transition matrix P is *reversible* if there is a probability distribution $\mathbf{u} \in \mathbb{R}^n$ such that for all $1 \leq i, j \leq n$,

$$u_i P_{i,j} = u_j P_{j,i}.$$

(a) If M is ergodic and reversible, then $\mathbf{u}$ is a stationary distribution of M.
(b) A random walk on a non-bipartite, connected graph is a reversible Markov chain.

(11) Compute the PageRank vector of the graph in Figure 5.1 with $c = 0.15$, $c = 0.5$, and $c = 1$. **Hint**: Use the power method.

(12) Compute the PageRank vector of the directed tree T in Figure 5.5 with $c = 0.15$. Interpret your results in light of the relationship between in-degree and PageRank.

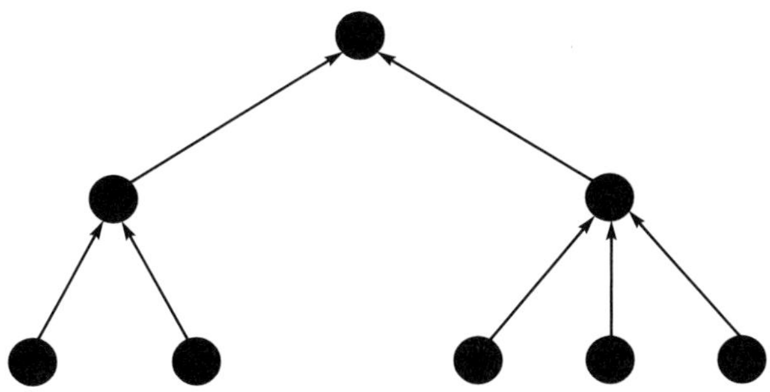

Figure 5.5. The directed tree T.

(13) Show that if we replace in P_2 the vector $1/nJ_{1,n}$ by an arbitrary stochastic *personalization vector* $\mathbf{v} > 0$, then the resulting PageRank Markov chain converges to a stationary distribution.

(14) If the eigenvalues of P_2 are $\{1, \lambda_1, \ldots, \lambda_n\}$, then prove that the eigenvalues of the PageRank matrix are $\{1, c\lambda_1, \ldots, c\lambda_n\}$.

(15) (a) Compute the HITS authority weights of vertices in the graph in Figure 5.3.

(b) Compute the SALSA hub and authority weights for the graph of (a).

(16) (a) Using the notation from the SALSA algorithm, verify that

$$\overline{A} = A_c^T A_r, \qquad H = A_r A_c^T.$$

(b) Show that the matrices $\overline{A}$ and H are stochastic.

(17) (a) Give an example of a digraph G so that the HITS authority vector for G depends on the seed vector.

(b) Give infinitely many examples of digraphs G so that the HITS authority vector for G depends on the seed vector.

Chapter 6

The Infinite Web

The infinite! No other question has ever moved so profoundly the spirit of man.

–David Hilbert

Nothing in life is to be feared. It is only to be understood.

–Marie Skłodowska-Curie

6.1. Introduction

While it possesses many billions of vertices and edges, the graph W is a *finite* graph. Or is it? A *static* web page provides the same downloadable information to all users, while the remaining pages are *dynamic*. The present number of static web pages is estimated to be around 54 billion [**132**]. The situation for the number of dynamic web pages is quite different. For example, type an arbitrary string s into a search engine to obtain a dynamic page P_s. There are infinitely many choices for s, and hence, infinitely many dynamic pages P_s. For another example, take an on-line calendar. The number of dates, and hence, the set of dynamic web pages is infinite (although practically finite owing to memory and other constraints).

Regardless of the finiteness of the universe, the view of W as an infinite graph presents an interesting perspective for the mathematician. Infinite graphs are fascinating creatures whose properties are often bewildering and quite unlike finite ones. As an example of the unusual behaviour found in infinite graphs, consider the infinite star-like graph G in Figure 6.1. To form G, attach a path of each finite length to a root vertex. This graph has the

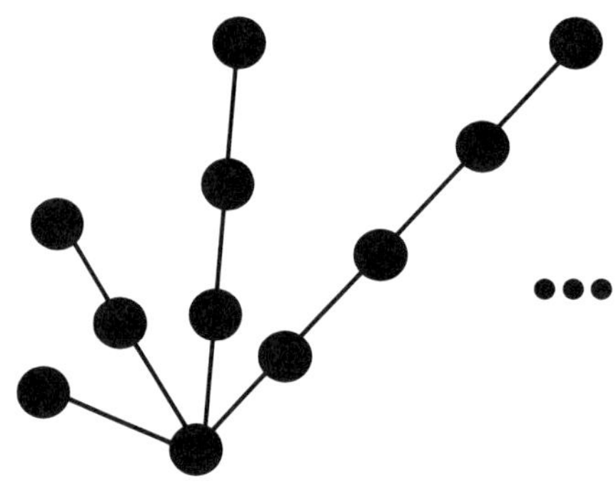

Figure 6.1. The infinite tree G.

property that there are uncountably many graphs H such that $G \leq H$ and $H \leq G$. (See Exercise 2.) We say that G and H are *mutually embeddable*; clearly the only graph mutually embeddable with a finite graph is itself.

Why study infinite graphs in relation to W? As we discussed in Chapter 1, the order and size of W are vast enough to pose real computational problems for the web programmer, and certainly there are more web pages out there than a search engine can catalogue, at least at the present time [**125**]. Further, the growth of the web makes its exact order hard to pin down.

Another reason to study infinite graphs in the context of W arises from the models we discussed in Chapter 4. These models generate finite graphs, but almost all of them are *on-line*: vertices are born over time. As time tends to infinity, the number of vertices grows without bound. In the actual limit, an infinite graph is born. These infinite on-line graph processes are not new. Deterministic versions of on-line graph processes were studied by Fraïssé [**111**] and others for many years prior to birth of the web.

If we consider a limit G of a chain of finite graphs generated by an on-line model, then the graph G can magnify latent properties of graphs generated by the model. For example, the robustness observed in the web graph can assert itself in G as a kind of *self-similarity property* in limits of web graph models: the graph G will contain many copies of itself, and making small changes in G will not change its isomorphism type. Infinite graphs are the natural playground for these kind of fractal properties, many of which are false for all finite graphs. Their presence in limits reflects back properties of the model, and hence, can be used as a diagnostic tool to distinguish models.

The approach of using infinite systems to study large finite ones is not specific to mathematics. This approach has been used successfully in physics (lattice structures), and in economics (continuum of agents). There is ample evidence to think that the infinite approach can glean many new insights into the mathematics of the web graph.

For an intriguing analogy, some cosmological models (such as the FLRW model [**93**]) are *homogeneous* (large regions of the universe have similar structure), while other models are *inhomogeneous*. For our graph models, we will find something similar: limits of $G(n, p)$ graphs are, with high probability, homogeneous (in a precise graph-theoretic sense) and satisfy a self-similarity property called *isotropic* (see Theorems 6.9 and 6.10). As we will see, while limits generated by web graph models are inhomogeneous in general, they are nevertheless isotropic (see Theorems 6.17, 6.29, and 6.32).

Infinite graphs possess their own rich theory which is quite different in many aspects from the finite case. In this chapter we will discuss some of these aspects, then attempt to apply infinite graphs to web graph theory. The approach is new and different, but we think it has and will lead to interesting mathematics and to better insights into W and its models. As is often the case in internet mathematics, results are more suggestive than definitive. Further, unlike previous chapters, the graphs we consider are largely deterministic, thus, the results presented are usually structural in nature.

In Section 6.2 we introduce the infinite random graph R, which is a prototype for the limits in later sections. In Section 6.4 we consider limits of copying models, while in Section 6.5, results on limits of preferential attachment models are presented. The final section deals with the new class of n-ordered graphs and their relation to limits. In each case, we will see that the limits have interesting structural properties (such as being isotropic) in common with R, but have other properties that distinguish them from R.

All the graphs we consider will be countable. The reason for this is that on-line web models generate finite graphs whose limits are countable. Hence, we may use the rich tools of finite graph theory to explore their properties. We denote the first infinite cardinal (which is the cardinality of $\mathbb{N}$) by $\aleph_0$, the first of the aleph cardinals $\aleph_\alpha$, where α is an ordinal.

Uncountable graph theory is interesting, but we rapidly encounter set-theoretic hurdles and questions there whose answer depends on the axioms of set theory we accept. As one example of this, it is independent of the ZFC axioms whether there exists a graph G of order $\aleph_2$ such that every subgraph of cardinality $\aleph_1$ of G has chromatic number less than or equal to $\aleph_0$ (see [**127**]).

6.2. The Infinite Random Graph

Consider the probability space $G(\mathbb{N}, p)$ consisting of graphs with $V = \mathbb{N}$ in which every distinct pair of integers is joined independently with probability p. The space $G(\mathbb{N}, p)$ is an infinite analogue of the finite random graph $G(n, p)$. Following our abuse of language from Chapter 3 (which, as we will see, is not as grievous as in the finite case!), we will call this space the *infinite random graph*. A surprising fact discovered by Erdős and Rényi is the following.

THEOREM 6.1. *With probability* 1, *all* $G \in G(\mathbb{N}, p)$ *are isomorphic.*

From one viewpoint, Theorem 6.1 makes the space $G(\mathbb{N}, p)$ uninteresting. Although this is true in a certain sense, the resulting unique isomorphism type, which we name R, has fascinated many researchers from several disciplines within mathematics, such as graph theorists, logicians, algebraists, and topologists. The proof of Theorem 6.1, while elementary, requires several tools: limits of finite graphs, an adjacency property, and an intriguing method to prove that graphs are isomorphic called *back-and-forth*.

There is an analogue of Theorem 6.1 using Baire category instead of probability. We may define a natural complete metric space on the set of all countable graphs by considering the edge set as a zero-one sequence. It can be shown that the set of graphs isomorphic to R is a residual subset. See [**60**].

Define a graph R^* as follows. Let R_0 be a K_1. Assume that for a non-negative integer $t \geq 0$, the graph R_t is defined and finite. To form R_{t+1} do the following. For each subset $S \subseteq V(R_t)$ (possibly empty) add a vertex z_S joined only to the vertices of S. Note that $|V(R_{t+1})| = |V(R_t)| + 2^{|V(R_t)|}$.

The sets $\{V(R_t) : t \in \mathbb{N}\}$ and $\{E(R_t) : t \in \mathbb{N}\}$ are well-ordered sets or *chains*, so we may legally define

$$V(R^*) = \bigcup_{t \in \mathbb{N}} V(R_t), \qquad E(R^*) = \bigcup_{t \in \mathbb{N}} E(R_t).$$

We write $\lim_{t \to \infty} R_t = R^*$, and say that R^* is the *limit of the chain* $(R_t : t \in \mathbb{N})$. The notion of limit extends to any chain $(G_t : t \in \mathbb{N})$ of finite or countable graphs. By taking the unions to start at some fixed $t_0 > 0$, we may extend this notion to limits of the form $\lim_{n \geq t_0} G_t$.

We introduce some notation that will simplify our discussion of graph limits. Let $J = \lim_{t \to \infty} H_t$ be a limit of a chain $\mathcal{C} = (H_t : t \in \mathbb{N})$ of graphs, where $H_t \leq H_{t+1}$ for all $t \in \mathbb{N}$. Define $\mathrm{age}_{\mathcal{C}} : V(J) \to \mathbb{N}$ by

$$\mathrm{age}_{\mathcal{C}}(x) = \begin{cases} t & \text{if } x \in V(H_t) \backslash V(H_{t-1}) \text{ where } t > 0; \\ 0 & \text{else.} \end{cases}$$

We will simply write age(x) if $\mathcal{C}$ is clear from context. The age of a finite subset, written age(S), is $\max\{\text{age}(x) : x \in S\}$.

The graph R^* has the following singular property. A graph G is *existentially closed* or *e.c.* if for all disjoint finite sets of vertices A and B (one of which may be empty), there is a vertex z joined to all of A and to no vertex of B. The mindful reader will realize this is just the conjunction of all the n-e.c. properties for $n \in \mathbb{N}$, which we introduced in Chapter 3. Following the notation from Chapter 3, we say that z is *correctly joined to* A and B. See Figure 6.2 for a depiction of the e.c. property.

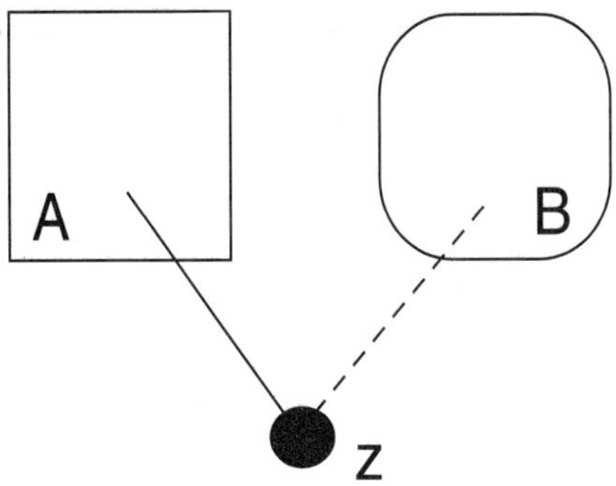

Figure 6.2. The e.c. property.

The notion of an existentially closed graph (or more generally, a first-order existentially closed structure) was introduced by the logician Abraham Robinson in the 1960's. Existential closure may be thought of as a generalization of algebraic closure in field theory. For more on existentially closed structures and their relationship to logic and model theory, see [**133**].

THEOREM 6.2. *The graph* R^* *is e.c.*

Proof. Fix finite disjoint A and B in $V(R^*)$. Let $t_0 = \text{age}(A \cup B)$. The vertex $z = z_A$ in R_{t_0+1} is correctly joined to A and B. □

We may think of $G(\mathbb{N}, p)$ as the limit of an on-line random graph process. At time 0, let X_0 be K_1. Assuming a graph X_t at time t is defined and finite, at time $t+1$, add a new vertex z to form X_{t+1}. The vertex z is joined to each existing vertex independently with probability p. Then $G(\mathbb{N}, p)$ consists of limits $\lim_{t\to\infty} X_t$. Graphs in $G(\mathbb{N}, p)$ are e.c. with high probability.

THEOREM 6.3. *With probability* 1, $G \in G(\mathbb{N}, p)$ *is e.c.*

Proof. Fix disjoint sets of vertices A and B in G with $|A| = i$ and $|B| = j$. For a given $z \notin A \cup B$, the probability that z is not correctly joined to A

and B is

$$1 - p^i q^j.$$

The probability that no vertex of G is correctly joined to A and B is therefore

$$\lim_{t\to\infty} \left(1 - p^i q^j\right)^{t-(i+j)} = 0.$$

As there are only countably many choices of A and B, and the countable union of probability 0 events has probability 0, the assertion follows. □

The following theorem and Theorem 6.3 together prove Theorem 6.1.

THEOREM 6.4. *If G and H are e.c. graphs, then $G \cong H$.*

Proof. The idea is to generate a chain of partial isomorphisms f_n between induced subgraphs of G and H, so that the limit of the chain F is an isomorphism of G with H. The mappings f_n are defined inductively, and care must be taken in the induction step to increase the orders of the corresponding domain and ranges.

Let $V(G) = \{x_i : i \in \mathbb{N}\}$ and $V(H) = \{y_j : j \in \mathbb{N}\}$. Define the mapping f_0 to be the isomorphism from $\langle x_0 \rangle_G$ to $\langle y_0 \rangle_H$.

For an induction hypothesis, suppose that f_t is defined for some $t \geq 0$, and is an isomorphism. If t is even, then suppose that f_t has domain containing $\{x_0, \ldots, x_t\}$, and if t is odd, f_t has range containing $\{y_0, \ldots, y_t\}$.

We consider cases depending on whether $t+1$ is even or odd. If $t+1$ is even, then we "go forward". Let x be the vertex of least index not in the domain of f_t. By induction hypothesis, the vertex $x \notin \{x_0, \ldots, x_t\}$ and so it could be x_{t+1}. We now define a mapping f_{t+1} whose domain is the domain of f_t and x (which ensures that x_{t+1} is in the domain of f_{t+1}). Suppose that the neighbours of x in the domain of f_t form a (finite) set S.

In H, by the e.c. property there is a vertex y joined to $f_t(S)$ and to no other vertex of the (finite) range of f_t. Define f_{t+1} to be the mapping extending f_t and mapping x to y. It is straightforward to see that f_{t+1} preserves adjacencies and non-adjacencies by the choice of y. Further, it is a bijection with the desired domain.

If $t+1$ is odd, then we "go back". Let y be the vertex of least index not in the range of f_t. By induction hypothesis, the vertex $y \notin \{y_0, \ldots, y_t\}$. Suppose that the neighbours of y in the range of f_t form a set T.

In G, there is a vertex x joined to $f_t^{-1}(T)$ and to no other vertex of the domain of f_t. Define f_{t+1} to be the mapping extending f_t and mapping x to y. As in the case of going forward, f_{t+1} is an isomorphism satisfying the conditions of the induction hypothesis.

For all integers $t \geq 1$ we have that

(1) If t is even, then $\{x_0, \ldots, x_t\}$ is in the domain of f_t.

(2) If t is odd, then $\{y_0, \ldots, y_t\}$ is in the range of f_t.

(3) f_{t+1} extends f_t.

(4) f_t is an isomorphism.

Define the mapping $F : G \to H$ by $F = \bigcup_{t \in \mathbb{N}} f_t$. More explicitly, $F(x) = f_i(x)$, where i is the least index such that $x \in V(R_i)$. Then F is well defined by item (3), and is a bijection by (1) and (2) since each f_t is an injection. It is an isomorphism by (4). To see this, suppose that ab is an edge of G. Choose $t \geq 0$ large enough so that a and b are in the domain of f_t (which is possible by (1)). Now $f(a)f(b) = f_t(a)f_t(b)$ is an edge of H by (4). The proof that non-edges are preserved is similar. □

The proof of Theorem 6.4 is often called the *back-and-forth* method, used historically first by Cantor in his proof that the rationals are the unique isomorphism type of countable linear order which is dense with no endpoints [**64**]. The technique was made popular by Fraïssé [**111**]. Back-and-forth is now a standard method of showing that two countably infinite graphs (or more general relational structures; see [**111**] and [**133**]) are isomorphic.

Back-and-forth can be phrased as a game of perfect information (that is, the players see and remember each other's moves) between two players on two graphs G_1 and G_2. The players are the *duplicator*, who is trying to show that G_1 and G_2 are isomorphic, and the *spoiler*, who is trying to prove the opposite. The game is played over countably many discrete time-steps (or rounds), with the spoiler beginning the play. During each round, each player chooses a vertex from either G_1 or G_2; however, if the spoiler chooses a vertex from G_i, then the duplicator must choose from $G_{i'}$, where $i' = 0$ if $i = 1$, and $i' = 1$, otherwise. This is called the *back-and-forth* or *Ehrenfeucht-Fraïssé* game.

The duplicator wins if he has a strategy so that at every round, the subgraphs induced by all the vertices chosen from G_1 and G_2 are isomorphic; the spoiler wins otherwise. The reader should verify that G_1 and G_2 are isomorphic if and only if for every play of the spoiler, the duplicator has a winning strategy. Back-and-forth games have connections to several topics such as expressibility in first-order and other logics. For further reading on back-and-forth games, the reader is directed to [**133**].

6.3. Representations and Properties of R

The graph R^* in the last section, which we now identify with R thanks to Theorem 6.4, has a simple definition. The main idea is to "extend" all finite sets in all ways over time. We call this a *representation of* R; that is, a graph isomorphic to R. The representation R^* has a similar flavour to our on-line models: over time new vertices are born and joined to existing ones via certain rules. As we will see, limits of on-line models are not usually e.c. (in fact, not even 1-e.c. in general). Note that we may start with any finite graph G as R_0 and the resulting limit will be isomorphic to R.

We now discuss two simple representations of R, both in terms of elementary number theory. The first representation of R uses binary expansions of integers, and was first given by Rado [**176**] (this is why R is often called the *Rado graph*). Define a graph R_1 with vertices $\mathbb{N}$. A vertex m is joined to n if 2^m occurs in the (unique) base 2 expansion of n, or if 2^n is in the base 2 expansion of m. For example 0 is joined to 1, and 4 is joined to 16, but 3 is not joined to 6.

THEOREM 6.5. *The graph R_1 is e.c., and hence, is isomorphic to R.*

Proof. Fix finite disjoint sets $A = \{i_1, \ldots, i_m\}$ and $B = \{j_1, \ldots, j_n\}$ of integers with $|A| = m$ and $|B| = n$. Without loss of generality, suppose that $i_1 < \cdots < i_m$ and $j_1 < \cdots < j_n$. Define

$$z = 2^{i_1} + \cdots + 2^{i_m} + 2^{j_n+1}.$$

By definition, z is joined to each vertex of A. To see that z is joined to no vertex of B, note that for all $1 \leq k \leq n$, 2^{j_k} is not in the base 2 expansion of z, and $j_k < z$. Theorem 6.4 finishes the proof. □

The second representation R_2 of R uses a few tools from number theory. Let the vertices of R_2 be the set of primes $\mathbb{P}_1$ congruent to 1 (mod 4). The set $\mathbb{P}_1$ is infinite by Dirichlet's theorem on primes in arithmetic progressions. Two distinct primes p and q in $\mathbb{P}_1$ are joined if p is a square (mod q) or q is a square (mod p). For example, 5 and 41 are joined, but 5 and 13 are not. The graph R_2 is undirected by the law of quadratic reciprocity, and recalls the Paley graphs defined in Chapter 3.

THEOREM 6.6. *The graph R_2 is e.c., and hence, is isomorphic to R.*

Proof. Fix finite disjoint sets $A = \{p_1, \ldots, p_m\}$ and $B = \{q_1, \ldots, q_n\}$ in $\mathbb{P}_1$ with $|A| = m$ and $|B| = n$. Let b_j be a non-square (mod q_j), where $1 \leq j \leq n$. By the Chinese remainder theorem, the following system of congruences has a solution $(\text{mod } 4\prod_{i=1}^{m} p_i \prod_{j=1}^{n} q_j)$:

$$\begin{aligned} z &\equiv 1 \pmod 4, \\ z &\equiv 1 \pmod{p_i}, \quad 1 \leq i \leq m, \\ z &\equiv b_j \pmod{q_j}, \quad 1 \leq j \leq n. \end{aligned}$$

Dirchlet's theorem on the arithmetic progression of primes asserts that there is a prime solution to this system of congruences. Hence, there is a z in $\mathbb{P}_1$ correctly joined to A, B. The assertion follows from Theorem 6.4. □

Other representations of R exist, a few of which are outlined in the exercises (see Exercises 3, 4, and 5). The graph R carries a rich structure, which has intrigued researchers for decades. For example, the following result is obtained by a variation of back-and-forth.

THEOREM 6.7. *The graph R is* universal*: all countable graphs embed in R.*

Proof. Let G be a fixed countable graph, and let $V(G) = \{x_i : i \in \mathbb{N}\}$. We name the subgraph induced by $\{x_i : 0 \leq i \leq t\}$ as G_t. Hence, $G = \lim_{t\to\infty} G_t$. We embed each G_t into R by induction, so each embedding extends the previous. To accomplish this, we go "forth" only. More precisely, let $f_0 : G_0 \to R$ be any fixed embedding. Suppose that for an integer $t \geq 0$, there is an embedding $f_t : G_n \to R$ extending f_0. The vertex x_{t+1} is joined to some set S of vertices in G_t. By the e.c. property, there is a vertex z of R joined only to $f_t(S)$ in G_t. Define the mapping $f_{t+1} : G_{t+1} \to R$ by extending f_t so that x_{t+1} is mapped onto z. The map f_{t+1} is an embedding by the choice of z. The mapping $F = \lim_{t\to\infty} f_t$ is an embedding of G into R. □

Another key property of R is its symmetry. One way of measuring symmetry is by automorphisms; the richer the automorphism group, the more symmetry the graph possesses. A graph G is *homogeneous* if each partial isomorphism between finite induced subgraphs extends to an automorphism of G. For example, each clique is homogeneous, as is the 5-cycle. We may view homogeneity as the strongest form of symmetry a graph can possess; for example, a homogeneous graph is both vertex- and arc-transitive.

The graph R is homogeneous by a back-and-forth argument (see Exercise 8). Homogeneous graphs are rare (as one might suspect) as is demonstrated by the following classification results. The classification of finite homogeneous graphs was completed independently by Gardiner [**114**], Gol'fand and Klin [**118**], and Sheehan [**185**].

THEOREM 6.8 ([**114, 118, 185**]). *A finite homogeneous graph is a disjoint union of complete graphs, a complete multipartite graph, C_5, or the Paley graph P_9.*

For an integer $n \geq 3$, the *Henson graph* H_n is defined as the limit graph formed by the following extension principle: for each (K_{n-1})-free induced subgraph S, add a vertex joined to S. For example, if $n = 3$, we extend independent sets, resulting in a triangle-free graph. Note that H_n is a K_n-free variant of the representation R^* of R. It is not hard to see (and you should check) that H_n is universal for all K_n-free graphs and is homogeneous. These graphs were first discovered and studied by Henson [**130**]. Lachlan and Woodrow [**148**] classified the countably infinite homogeneous graphs.

THEOREM 6.9 ([**148**]). *The countably infinite homogeneous graphs are, up to isomorphism, the following.*

(1) *The graphs mK_n (m disjoint copies of complete graphs of size n), where m and n are cardinals with the property that $m + n = \aleph_0$.*
(2) *The complements of graphs of (1).*
(3) *The Henson graphs H_n, $n \geq 3$.*
(4) *The complements of graphs of (3).*
(5) *The graph R.*

A Polish space is a complete separable metric space. A Polish space is *homogeneous* if every isometry between finite subspaces extends to an isometry of the whole space, and is *universal* if every Polish space isometrically embeds in it. It is interesting to note that in 1927, a paper of P. Urysohn constructed a unique isomorphism type of universal homogeneous Polish space (albeit the space is uncountable). The discovery of what is now called the *Urysohn space* predates the work of Fraïssé on homogeneous relational structures by two decades. See [**62**] for more on the Urysohn space and its relationship to R.

The infinite random graph has a rich structure. A property of graphs with the flavour of self-similarity in fractals is the following. A limit $G = \lim_{t\to\infty} G_t$ is *isotropic* if for all $t_0 \geq 0$, there is an $H \leq \lim_{t\geq t_0} G_t$ such that $H \cong G$. In other words, G is isotropic if at any given time t_0, we may find an induced subgraph contained in the limit $\lim_{t\geq t_0} G_t$ that is isomorphic to G. For example, the infinite one-way path P considered as the limit of the sequence $(P_t : t \in \mathbb{N})$ is isotropic. To see this, it is not hard to verify that for all $t_0 \geq 0$, each of the graphs $\lim_{t\geq t_0} P_t$ is isomorphic to P. However, there are $2^{\aleph_0}$ many non-isomorphic graphs which are not isotropic (we could call these *anisotropic*). For example, the graph G which is the limit of the sequence of graphs $C_3 \uplus \cdots \uplus C_t$, where $t \geq 3$, is anisotropic.

The infinite random graph satisfies something much stronger than the isotropic property. We say that a graph G has the *pigeonhole property*, written $(\mathcal{P})$, if for each 2-colouring (that is, a partition of $V(G)$ into two parts) of the vertices of G, the induced subgraph with vertices of the same colour is isomorphic to G. For example, it is easy to see (by the pigeonhole property for sets) that the infinite clique $K_{\aleph_0}$ and its complement $\overline{K_{\aleph_0}}$ have

$(\mathcal{P})$. As one application of $(\mathcal{P})$, for all $x \in V(R)$, the subgraph induced by $N(x)$ is isomorphic to R (since $V(R) \backslash N(x)$ has an isolated vertex and R is connected of diameter 2 by the e.c. property).

The pigeonhole property is in fact fairly rare, as the following theorem of Peter Cameron [**60**] demonstrates. It represents another characterization of R.

THEOREM 6.10. *The graphs with $(\mathcal{P})$ are, up to isomorphism, K_1, $K_{\aleph_0}$, $\overline{K_{\aleph_0}}$, and R. In particular, R is the unique* 1*-e.c. graph with $(\mathcal{P})$.*

Proof. The graph K_1 vacuously satisfies $(\mathcal{P})$, and we have already discussed the graphs $K_{\aleph_0}$, $\overline{K_{\aleph_0}}$. It is an exercise that R satisfies $(\mathcal{P})$ (Exercise 12).

Now suppose that G is a graph satisfying $(\mathcal{P})$ that is neither a clique nor the complement of a clique. Suppose that G contains isolated vertices; let X be the set of all the isolated vertices of G, and let $Y = V(G) \backslash X$. By $(\mathcal{P})$, $\langle Y \rangle \cong G$. But this is a contradiction, since $\langle Y \rangle$ has no isolated vertices. Since the graph $\overline{G}$ also has $(\mathcal{P})$ (verify this), we have therefore proved that G has neither universal nor isolated vertices. Hence, G is 1-e.c.

For a fixed integer $n \geq 1$, assume as the induction hypothesis that G is n-e.c. We now prove that G is $(n+1)$-e.c.

Fix disjoint sets A and B of vertices such that $|A \cup B| = n + 1$. By considering $\overline{G}$, without loss of generality, we may assume that A is non-empty. Say $x \in A$, and let $A' = A \backslash \{x\}$. By the induction hypothesis, there is a vertex correctly joined to A', B. Let X be the set of all vertices of G correctly joined to A', B, and let Y be the remaining vertices of G. Although $A', B \subseteq Y$, there is no vertex of $\langle Y \rangle$ correctly joined to A', B. By $(\mathcal{P})$, we must have that $\langle X \rangle \cong G$. Hence, there are infinitely many vertices correctly joined to A', B.

Now consider $X' = X \cup \{x\}$ and Y' the remaining vertices. By a similar argument as above, we must have that $\langle X' \rangle \cong G$. As G is 1-e.c., $N(x)$ is non-empty in X'. But then a vertex z of $N(x)$ in X' is correctly joined to A and B in G. Hence, G is e.c., and so is isomorphic to R by Theorem 6.4. □

The isotropic property follows from $(\mathcal{P})$; however, there are $2^{\aleph_0}$ many non-isomorphic isotropic graphs. (See Exercise 13.) The pigeonhole property straightforwardly generalizes to any relational structure. Hence, the classification problem of structures with $(\mathcal{P})$ becomes a natural problem. The countable tournaments and orders with $(\mathcal{P})$ were classified in [**43**], while the digraphs with $(\mathcal{P})$ were classified in [**85**]. For $k \geq 2$, a *k-uniform hypergraph* is a one where each hyperedge has cardinality k. For example, a

2-uniform hypergraph is a graph. It is an open problem to classify the isomorphism types of k-uniform hypergraphs for $k \geq 3$ with $(\mathcal{P})$.

6.4. Limits of Copying Models

In the web graph, each vertex acts as an independent agent, which will base its decision on how to link to the existing network on local knowledge. As a result, the neighbourhood of a new vertex will often be an imperfect copy of the neighbourhood of an existing vertex. This aspect of web page generation indicates a weakness of preferential attachment models: we assume global knowledge of all vertex degrees, a clearly unrealistic hypothesis when faced with a massive set of vertices. Both the *copying models* [**3, 145**] of the web graph, and the *duplication model* [**69**] for biological networks incorporate this notion of copying in their definitions.

As in the case with finite web graph models, the probability space $G(\mathbb{N}, p)$ is not suitable for our purposes. Instead we introduce a generalized copying model from [**48**], which in a certain sense generalizes and unifies the copying and duplication models. The three parameters of the model $\text{Copy}(p, \rho, H)$ are $p \in [0, 1)$, a function $\rho : \mathbb{N} \to \mathbb{N}$, and a fixed finite initial graph H. To simplify the discussion, we require that $\rho(t)$ is an integer-valued, non-decreasing function such that, for some $\alpha < 1$, $\rho(t) \leq \alpha t$ for all $t \geq t_0$. Moreover, unless otherwise stated we assume that $\rho(t) = \Theta(t^s)$ for some $s \in [0, 1]$. The model describes a random graph process over a countable sequence of discrete time-steps indexed by $t \in \mathbb{N}$.

(1) At $t = 0$, set $G_0 = H$.

(2) For a fixed $t > 0$, assume that G_{t-1} has been defined, has order $t - 1 + |V(H)|$, and contains G_0 as an induced subgraph. To form G_t, add a vertex v_t to G_{t-1} and choose its neighbours as follows.

(3) Choose an existing vertex u_t from G_{t-1} u.a.r.
   - (a) For each neighbour w of u_t, independently add an edge from v_t to w with probability p.
   - (b) Choose $\rho(t)$ many vertices from $V(G_{t-1})$ u.a.r., and add edges from v_t to each of these vertices. Call these *extra* edges.

We note that if $\rho(t) = 0$, then the graphs G_t generated by the model $\text{Copy}(p, \rho, H)$ correspond exactly to the graphs generated by the duplication model. If $\rho(t)$ is constant and $p > 0$, then the graphs G_t are undirected analogues of graphs generated by the copying model.

A new adjacency property introduced in the context of limits of copying models graphs by Bonato and Janssen [**46**] is a local version of the existentially closed property. A graph G is *locally e.c.* (originally given the less descriptive name (B) in [**46**]) if for each vertex y of G, for each finite

$X \subseteq N(y)$, and each finite $Y \subseteq V(G) \setminus X$, there exists a vertex $z \neq y$ which is correctly joined to X and Y. The locally e.c. property is a variant of the

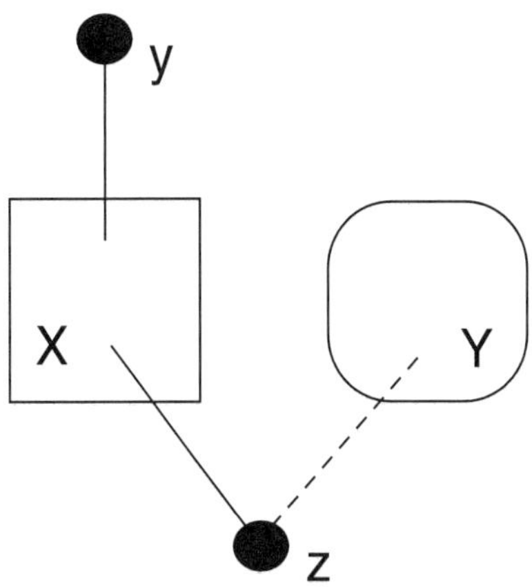

Figure 6.3. The locally e.c. property.

e.c. property that applies only to sets X contained in the neighbour set of a vertex. In simple language, we can extend all neighbour sets in all possible ways. The model Copy(p, ρ, H) is very likely to generate limits satisfying the locally e.c. property, following the proof of Theorem 2.1 in [**46**]. The proof below has similar flavour to our proof of Theorem 6.1, but requires more care.

THEOREM 6.11. *Fix $p \in (0,1)$, H a finite graph, and let ρ be a proper function. With probability* 1, *a limit $G = \lim_{t\to\infty} G_t$ of graphs generated by the model* Copy(p, ρ, H) *is locally e.c.*

Proof. Since a countable union of measure 0 subsets has measure 0, it suffices to show that for a fixed $y \in V(G)$, and finite disjoint $X \subseteq N(y), Y \subseteq V(G)$, the probability that there is no vertex correctly joined to all of X, Y is 0 (since there are only countably many choices for y and X, Y in G).

Fix a vertex y and disjoint sets $X \subseteq N(y)$ and Y in $V(G)$, and let $t_0 = \text{age}(X \cup Y \cup \{y\})$. Let $Y_1 = Y \cap N(y)$, and let $|X| = k$, $|Y| = j$, and $|Y_1| = j_1$. Let u_t be the copy vertex in time-step t, and let $B_{X,Y}(t)$ be the event that u_t is correctly joined to X and Y. Let $B'_{X,Y}(t)$ be the event that $u_t = y$, every edge from y to vertices in X is copied, none of the edges from y to vertices in Y_1 are copied, and no vertex in Y receives an extra edge in time-step t. Note that $B'_{X,Y}(t)$ implies $B_{X,Y}(t)$. Also, the events $B'_{X,Y}(t)$ for different values of t are independent.

If $t > 0$, then the probability that $u_t = y$ is

$$\frac{1}{t-1+|V(H)|} = \Theta\left(\frac{1}{t-1}\right),$$

since u_t is chosen u.a.r. from G_{t-1}. The probability that all edges to X and no edges to Y are copied, given that $u_t = y$, equals $p^k(1-p)^{j_1}$. For all $t > 0$ such that $j < t - \rho(t)$, the probability that no vertex in Y receives an extra edge equals

$$\frac{\binom{t-j-1}{\rho(t)}}{\binom{t-1}{\rho(t)}} \geq \left(\frac{t-j-\rho(t)}{t-\rho(t)}\right)^j \geq \left(1-\frac{j}{t-\rho(t)}\right)^j. \tag{6.1}$$

As ρ is proper, there is a $t_1 \geq t_0$ and $\alpha \in (0,1)$ so that $\rho(t) \leq \alpha t$ for all $t \geq t_1$. Then by (6.1), the above probability is at least

$$f(t) = \left(1 - \frac{j}{(1-\alpha)t}\right)^{\alpha t}.$$

Note that $f(t)$ is an increasing function of t with range in $(0, 1)$. Therefore, for all $t \geq t_1$,

$$\mathbb{P}(B'_{X,Y}(t)) \geq \frac{1}{t-1} p^k(1-p)^{j_1} f(t_1).$$

Let $d = p^k(1-p)^{j_1} f(t_1) \in (0,1)$. Let C be the event that no vertex of G is correctly joined to X, Y. Then

$$\begin{aligned} \mathbb{P}(C) &\leq \mathbb{P}\left(\bigcap_{t=t_1}^{\infty} \overline{B'_{X,Y}(t)}\right) \\ &\leq \prod_{t=t_1}^{\infty}\left(1-\frac{d}{t-1}\right) = 0, \end{aligned}$$

by elementary properties of infinite products. □

Theorem 6.11 shows that the locally e.c. property plays an important role in generalized copying models. Besides their relationship to these models, locally e.c. graphs are of interest in their own right. Contrary to the situation for countable e.c. graphs (which are isomorphic to the infinite random graph R), not all countable locally e.c. graphs are isomorphic. In fact, there are $2^{\aleph_0}$ many non-isomorphic countable locally e.c. graphs (the maximum possible number of isomorphism types). This marks one large difference in the theories of e.c. and locally e.c. graphs.

The question of whether limits of graphs generated by copying models converge to R was addressed in [**47**]. For a non-negative integer n, a graph is *strongly n-e.c.* if for each pair of disjoint, finite sets of vertices A and B with $|A| = n$, there is a vertex z not in $A \cup B$ correctly joined to A and B. Hence, no restriction is put on $|B|$ in terms of n. Note that a graph G is *strongly* 0-*e.c.* if and only if for each finite set $B \subseteq V(G)$, there is a vertex not in B that is not joined to any vertex of B. For notational consistency, we say that a graph is *strongly* ∞-*e.c.* if it is e.c. If G is strongly n-e.c., then G is infinite and strongly m-e.c. for all positive $m < n$ (vertices not in A or B can be added to A to form a set of cardinality n). Further, the degree of each vertex in a strongly n-e.c. graph is infinite.

We describe a sharp threshold for the values of n for which an infinite limit generated by Copy(p, ρ, H) has the strongly n-e.c. property. In particular, for an extra edge function $\rho(t) = \Theta(t^s)$, we define a value $n_{p,s}$ below and show that any limit of Copy(p, ρ, H) is strongly n-e.c. if $n \leq n_{p,s}$, but, with positive probability, not n-e.c. if $n > n_{p,s}$.

For $p, s \in (0, 1)$, define the integer

$$n_{p,s} = \max\left(\left\lfloor \frac{1}{1-s} \right\rfloor, \lfloor \log_p(1-s) \rfloor + 1\right).$$

For all $s \in (0, 1)$, we define $n_{0,s} = \lfloor \frac{1}{1-s} \rfloor$, and for all $p \in (0, 1)$, $n_{p,1} = \infty$.

THEOREM 6.12 ([**47**]). *Let* $p \in [0, 1)$, $\rho(t) = \Theta(t^s)$ *for some* $s \in [0, 1]$, *and* H *be a finite graph. Let* $G = \lim_{t\to\infty} G_t$ *be generated by* Copy(p, ρ, H). *With probability* 1, G *is strongly* $(n_{p,s})$-*e.c. In particular, if* $s = 1$, *then* G *is isomorphic to* R.

In the case $p = 0$, no copying occurs and only $\rho(t)$ extra edges are added to the new vertex in each time-step. This includes the *growing m-out* model (see [**36**]) where $p = 0$ and $\rho(t) = m$. In the case that $p = 0$, we have a sharp threshold for the strongly n-e.c. property at $n_{0,s} = \lfloor \frac{1}{1-s} \rfloor$. For values of $p > 0$ and s such that $n_{p,s} = \lfloor \frac{1}{1-s} \rfloor$, the infinite limit behaves similarly to the case where $p = 0$. This provides additional evidence that in this case the copy behaviour plays a secondary role in the generating process.

On the other hand, we have the following theorem.

THEOREM 6.13 ([**47**]). *Let* $p \in [0, 1)$, $\rho(t) = \Theta(t^s)$ *for some* $s \in [0, 1)$, *and* H *be a finite graph. Let* $G = \lim_{t\to\infty} G_t$ *be generated by* Copy(p, ρ, H).

(1) *If* $\rho(t) = 0$, *then with positive probability,* G *is not* 1-*e.c.*

(2) *If* $s = 0$, *where* m *is a positive integer, then with positive probability,* G *is not* 2-*e.c.*

(3) *If* $s \in (0, 1)$, *then with positive probability,* G *is not* $(n_{p,s} + 1)$-*e.c.*

We now introduce the following limit graphs. Fix a finite graph H. Let $R_0 \cong H$. For a fixed $t \geq 0$, assume that R_t is defined and is finite. For each vertex $y \in V(R_t)$, and each subset $X \subseteq N(y)$, add a new vertex $z_{y,X}$ joined only to X. This gives the graph R_{t+1} which contains R_t as an induced subgraph. Define $R_H = \lim_{t\to\infty} R_t$. Hence, starting with H, at each time-step we extend locally in all possible ways. We call R_H a *copying graph.* Note that there is one copying graph for each finite H, and they are all locally e.c. (mimic the proof of Theorem 6.2).

We collect some elementary but useful facts about the copying graphs. We write $G \uplus H$ for the disjoint union of two graphs, and $\biguplus_{i=1}^{n} G_i$ for the disjoint union of graphs $G_1, \dots, G_n$.

LEMMA 6.14. *Fix a finite graph H.*

(1) *The degree of each vertex of R_H is $\aleph_0$ or 0.*
(2) $\chi(R_H) = \chi(H)$.
(3) $\omega(R_H) = \omega(H)$.
(4) *If H has connected components $\{H_i : 1 \leq i \leq n\}$, then*

$$R_H \cong \biguplus_{i=1}^{n} R_{H_i}.$$

(5) *Assume that H is connected and non-trivial. The graph R_H is isomorphic to $C_H \uplus \overline{K_{\aleph_0}}$, where C_H is locally e.c. and connected of infinite diameter.*

Proof. Suppose that a vertex x of R_H is not isolated, but has finite degree k. By the locally e.c. property, there is a vertex v joined to x but not joined to any vertex of $N(x)$. This contradiction proves (1).

For item (2), colour $H = R_0$ with $m = \chi(H)$ many colours. Assuming that $m = \chi(R_t)$ where $t \geq 0$ is fixed, colour $z_{y,X}$ by the colour of y. This gives a proper m-colouring of R_{t+1}. Hence, $\chi(R_H) = m$.

For (3), as $H \leq R_H$, we have that $\omega(H) \leq \omega(R_H)$. We prove by induction that $\omega(R_t) \leq \omega(H) = n$, thereby $\omega(R_H) \leq \omega(H)$. As the base case is immediate, assume that for a fixed $t \geq 0$, $\omega(R_t) \leq n$. Suppose for a contradiction that $\omega(R_{t+1}) = n + 1$ (note that $V(R_{t+1})\backslash V(R_t)$ is an independent set). Then there is a copy K of K_{n+1} with one vertex z with age $n+1$. Suppose that $z = z_{y,X}$ for some y, X with age$(\{y\} \cup X) \leq t$ and $X \subseteq N(y)$. But $\langle (V(K)\backslash\{z\}) \cup \{y\}\rangle$ is an $(n+1)$-clique with age at most t, which contradicts the induction hypothesis.

To prove item (4), note that if x and y are in different components in H, then they are never connected by a path in R_H. This is obtained by induction, with the base-step following immediately. Suppose that x and y are not connected at time t, where $t \geq 0$ is fixed. For a contradiction, suppose that x and y are joined by a path P in R_{t+1}. Then P must contain a unique vertex z with age $t+1$. Suppose that $z = z_{y,X}$ for some y, X with $\mathrm{age}(\{y\} \cup X) \leq t$ and $X \subseteq N(y)$. Then $(P \backslash \{z\}) \cup \{y\}$ is a path connecting x and y in R_t, which is a contradiction.

Given a connected component H_i of H, it is evident that the on-line process which generates R_H generates an induced subgraph which is isomorphic to R_{H_i}. By what was proved in the previous paragraph, all of these subgraphs are disjoint, and it is not hard to see that each vertex is in one of the R_{H_i}. Hence, $R_H \cong \biguplus_{i=1}^{n} R_{H_i}$.

For (5), first write $C_0 = R_0$. For each $t \geq 1$, write $R_t = C_t \uplus I_t$, where C_t is the subgraph induced by vertices joined to some vertex of older age, and $I_t = V(R_t) \backslash V(C_t)$. The graph C_t is connected by induction, and the set I_t consists of isolated vertices.

Notice that $|I_{t+1}| > |I_t|$ and $|I_1| > 0$, since isolated vertices are introduced at each time-step, and remain isolated in all later time-steps. Hence, for all t, $I_t \leq I_{t+1}$ and $C_t \leq C_{t+1}$. Define $C_H = \lim_{t\to\infty} C_t$ and $I_H = \lim_{t\to\infty} I_t$. It follows that $I_H \cong \overline{K_{\aleph_0}}$ and C_H is connected.

To show that C_H is locally e.c., fix a vertex y of C_H, and finite sets $X \subseteq N(y)$ and $Y \subseteq V(C_H) \setminus X$. If X is empty, then add an arbitrary vertex u from $N(y) \backslash X$ to X. By item (1) we may find such a u. In R_H we can find a vertex $z \neq y$ correctly joined to X and Y. But then z is in C_H as X is non-empty.

We now show that C_H has infinite diameter. We argue that there are paths of length at least k in C_H by induction on k. As H is connected and non-trivial, the induction begins. Suppose that x and y are distance at least $k \geq 1$ in C_H. Let P be a path of length at least k connecting x and y. By the locally e.c. property, there is a vertex z joined to y but to no other vertex of P. The path formed by adjoining z to P is of length at least $k+1$. □

Note that by Lemma 6.14, none of the graphs R_H are 1-e.c. As the graphs C_H are of infinite diameter, they are not 2-e.c. It is becoming evident to the reader that the graphs R_H behave quite differently than R.

THEOREM 6.15. *There are $2^{\aleph_0}$ many non-isomorphic infinite locally e.c. graphs.*

Proof. For a fixed integer $n \geq 0$, let $G = R_{K_n}$. By Lemma 6.14 (2), $\chi(G) = \chi(K_n) = n$, and $G = C_{K_n} \uplus \overline{K_{\aleph_0}}$. As C_{K_n} contains K_n, we have that $\chi(C_{K_n}) = n$.

Now let

$$\Omega = \{C_{K_n} : n \geq 4\},$$

which has cardinality $2^{\aleph_0}$. For a fixed $\varnothing \neq X \subseteq \Omega$, define $G(X)$ as $\biguplus_{G \in X} G$. Then $G(X)$ is locally e.c. To see this, fix $y \in V(G(X))$, and finite $X \subseteq N(y)$

and Y disjoint from X. If $y \in V(C_{K_n})$, then so is X. As C_{K_n} is locally e.c. by Lemma 6.14 (5), a vertex z joined to X and to no vertex of $Y \cap V(C_{K_n})$ is correctly joined to X and Y. If y is an isolated vertex, then X must be empty. As $X \cup Y$ is finite, any of the infinitely many isolated vertices not in $\{y\} \cup X \cup Y$ is correctly joined to X, Y.

Now if X, Y are subsets of Ω with $X \neq Y$, then suppose that $C_{K_n} \in X \backslash Y$. There is no component in $G(Y)$ with chromatic number n, so $G(X) \not\cong G(Y)$. Hence, there are $2^{\aleph_0}$ many non-isomorphic locally e.c. graphs. □

Despite Theorem 6.15 we observe, however, that the graphs R_H are *minimal* graphs with the given adjacency property, in the sense that they embed (as induced subgraphs) in any graph with the property.

THEOREM 6.16. *Fix a finite graph H. If G is a locally e.c. graph, then $R_H \leq G$ if and only if $H \leq G$.*

Proof. The forward direction is immediate, since $H \leq R_H$. For the reverse direction, suppose that H is an induced subgraph of G. We proceed by induction on t to show that each of the graphs R_t (used to define R_H at the beginning of this section) is an induced subgraph of G extending the embedding of H in G. We take R_0 to be this copy of H.

Fix a vertex $y \in V(R_t)$, and a finite subset $X \subseteq N(y)$. Let $Y = V(R_t)\backslash X$. As G is locally e.c., there is a vertex z correctly joined to X, Y. The vertex z is joined only to X in R_t, and plays the role of the vertex $z_{y,X}$ in the definition of R_H. □

Like R, the graphs R_H are isotropic.

COROLLARY 6.17. *For a fixed finite H, the graph R_H is isotropic.*

Proof. Fix an integer $t_0 > 0$. We use Theorem 6.16 to prove that R_H embeds in $G = \lim_{t \geq t_0} R_t$. To see that G is locally e.c., fix $y \in V(G)$, and finite subsets $X \subseteq N(y)$, $Y \subseteq V(G)$. Let $t' = \text{age}(\{y\} \cup X \cup Y)$. In R_H there is a vertex z correctly joined to X and $V(R_{t'})$. Hence, z is in G.

We show that $H \leq G$ to finish the proof. List the vertices of $R_0 \cong H$ as $\{1, 2, \ldots, k\}$. In R_{t_0+1}, there is a vertex $1'$ joined to R_0 in the same way as 1. Suppose by induction that there are vertices $1', \ldots, j'$ in G which are joined to each other and to R_0 in the same way as corresponding vertices in H. In particular, $\langle 1', \ldots, j' \rangle_G \cong \langle 1, \ldots, j \rangle_{R_0}$. Now the vertex $j+1$ is joined to some set S of vertices in H, and so is joined to a corresponding set S' in $\langle 1', \ldots, j' \rangle_G$. By the locally e.c. property in G, there is a vertex $(j+1)'$ correctly joined to $S \cup S'$ and $\{1', \ldots, j'\}$. Clearly,

$$\langle 1', \ldots, (j+1)' \rangle_G \cong \langle 1, \ldots, j+1 \rangle_H.$$

By induction, we obtain that

$$H \cong \langle 1', \dots, k' \rangle_G \leq G. \qquad \square$$

The following theorem establishes an intriguing connection between graph homomorphisms and the induced subgraphs of R_H.

THEOREM 6.18. *Fix a positive integer n, with G and H finite graphs. Then $G \leq R_H$ if and only if $G \to H$.*

Proof. For the forward direction, suppose that $G \leq R_H$. Then $G \to R_H$. The reader should verify that $R_H \to H$. Hence, $G \to H$ by the transitivity of the homomorphism relation.

For the converse, we introduce an auxiliary graph construction. Fix $f : G \to H$ a homomorphism. Assume that $V(G)$ and $V(H)$ are disjoint, and define a graph $H(G, f)$ to have vertices $V(G) \cup V(H)$, and edges

$$E(G) \cup E(H) \cup \{xy : x \in V(G), y \in V(H), \text{and } f(x)y \in E(H)\}.$$

We refer to the induced copy of H in $H(G, f)$ as H'. The graph $H(G, f)$ is just the union of G and H, so that for each vertex x of G, x is joined to all the neighbours of $f(x)$ in H. (See Figure 6.4 for an example.) We proceed by induction on $|V(G)|$ to show that $H(G, f) \leq R_H$. (Since $G \leq H(G, f)$, this proves the statement.) Note that, if $|V(G)| = 1$, then $H(G, f)$ is isomorphic to the disjoint union of H and K_1. The base case follows. The induction hypothesis is that if $|V(G)| = n$, where $n \geq 1$ is fixed, then $H(G, f)$ is an induced subgraph of R_H with H' the copy of H at $t = 0$.

Let $|V(G)| = n + 1$, and fix $x \in V(G)$. By the induction hypothesis, $H(G - x, f \restriction G - x)$ is a subgraph of R_H with H' the copy of H at $t = 0$. By the definition of $H(G, f)$, all the neighbours of x in $H(G, f)$ are also neighbours of the vertex $f(x)$ in H'. By the locally e.c. property of R_H, there is a vertex z of R_H joined exactly to the neighbours of x in $H(G, f)$. Adding z to the copy of $H(G - x, f \restriction G - x)$ in R_H will give an induced subgraph of R_H which is isomorphic to $H(G, f)$, while H' is unchanged. This completes the induction. $\square$

One interpretation of Theorem 6.18 is that the graph R_H carries a certain memory of H, made explicit by the homomorphism to H.

The proof of Theorem 6.18 can be modified if G is countable. Hence, we have the following corollary.

COROLLARY 6.19. *For a fixed finite graph H, all countable H-colourable graphs embed in R_H; that is, R_H is a universal H-colourable graph.*

For example, by Corollary 6.19, $R_{K_2} \leq R_{C_6}$ and $R_{C_6} \leq R_{K_2}$. However, $R_{K_2} \not\cong R_{C_6}$ (Exercise 14). A more suggestive application of Corollary 6.19

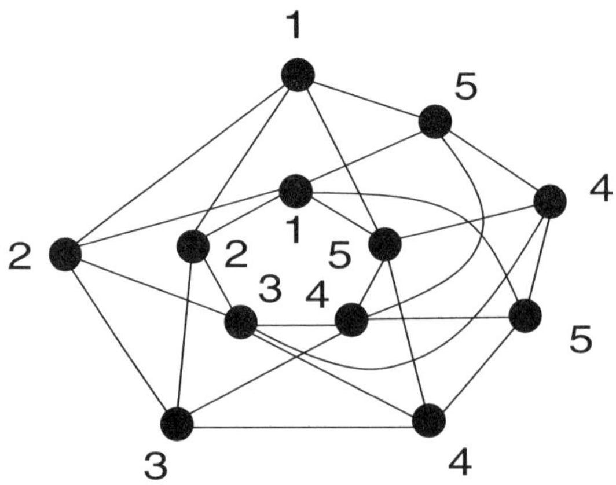

Figure 6.4. The graph $C_5(C_7, f)$, where the homomorphism f to C_5 is defined by the vertex labellings.

is that for all finite non-trivial H, all countable bipartite graphs embed in R_H. This is reminiscent of the community structure observed in the web graph, where a large number of bipartite cliques were reported in [**146**].

By Theorem 6.16 and Corollary 6.19, the graphs R_H are then both the maximal H-colourable graphs and the minimal locally e.c. graphs containing H (with respect to the embedding order on countable graphs). We will note a similar situation for the graphs $R^{(n)}$ introduced in the final section of this chapter.

From the classification of homogeneous graphs of Theorem 6.9, we deduce that R_H is homogeneous if and only if H is K_1 (since $R_{K_1} \cong \overline{K_{\aleph_0}}$). Hence, the graphs R_H are not nearly as symmetric as R. In fact, the graphs R_H, in general, are not even vertex-transitive.

LEMMA 6.20. *The graph R_{K_3} is not vertex-transitive.*

Proof. List the vertices of K_3 at time $t = 0$ as a, b, c. Consider the vertex $x = x_{a,\{b\}}$ added at $t = 1$. Therefore, x is joined to b but to neither a nor c. We claim that x is not in a K_3 in R_{K_3}. Hence, R_{K_3} is not vertex-transitive since there is no automorphism sending a to x. We proceed by induction on $t \geq 1$ to prove that x is not in a K_3 in R_t. This clearly holds for $t = 1$. Suppose it is true for a fixed $t \geq 1$, and consider R_{t+1}. For a contradiction, suppose that x is in a K_3 with vertices x, y, z. One of these vertices, say z, must have age $t + 1$. Let z' be the vertex z copied from at time $t + 1$. Note that $x, y \in N(z) \subseteq N(z')$. Therefore, $z' \neq x, y$. Then x, y, z' form a K_3 in R_t, which is a contradiction. $\square$

We note that it is an open problem to determine for which finite graphs H the copying graphs R_H are vertex-transitive.

A small variation on locally e.c. gives something equivalent to the e.c. property. We say that a graph G has the *n-locally e.c.* for some fixed positive $n \in \mathbb{N}$ if for each vertex y of G, for each finite $X \subseteq N(y)$, for each finite $Y \subseteq V(G) \setminus X$, and for each set $U \subseteq V(G) \setminus (N(y) \cup Y)$ with cardinality at most n, there is a vertex $z_{X,Y,U} \neq y$ correctly joined to $X \cup U$ and Y. See Figure 6.5

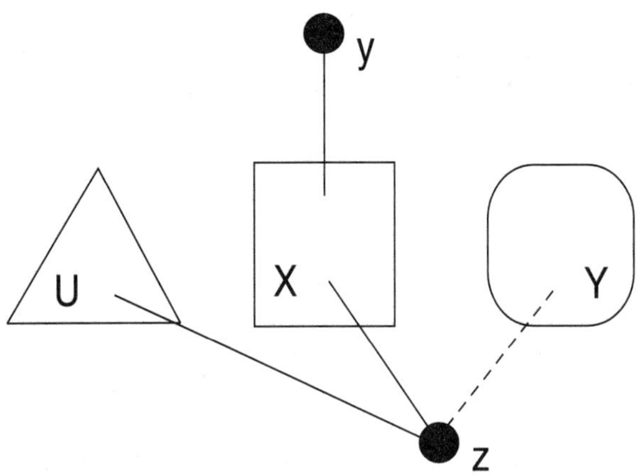

Figure 6.5. The n-locally e.c. property.

THEOREM 6.21. *Fix an integer $n > 0$. If G is n-locally e.c., then G is isomorphic to R.*

Proof. To show that G is isomorphic to R, we need only show that G is e.c.: for all finite disjoint subsets C and D of vertices of G, there is a vertex z of $V(G) \setminus (C \cup D)$ that is joined to each vertex of C and to no vertex of D.

As C is finite, we may write C as $W_1 \cup \cdots \cup W_r$, where the W_i are pairwise disjoint and have cardinality at most n. Choose a vertex y not in $C \cup D$. Let $X = N(y) \cap W_1$, $Y = D$, and let $U = W_1 \setminus N(y)$. By the n-locally e.c. property, there is a vertex $x_1 \notin (X \cup Y)$ such that $W_1 \subseteq N(x_1)$. By the n-locally e.c. property with $X = W_1$, Y empty, and $U = W_2$, there is a vertex x_2 not in $X \cup Y$ that is joined to all of $W_1 \cup W_2$. Proceeding inductively, we can find a vertex x_r that is joined to all of the vertices in C. As then $C \subseteq N(x_r)$, by a final application of the n-locally e.c. property, there is a vertex z correctly joined to C and D. $\square$

It is surprising that the seemingly weak 1-locally e.c. is equivalent to the e.c. property. The introduction of one "random" vertex (that is, a vertex from the set U) into the locally e.c. property reverts the whole picture back to R.

6.5. Limits of Preferential Attachment Models

Preferential attachment is one of the main paradigms used in the design of web graph models. Hence, it is natural to consider limits of on-line preferential attachment models. This approach was first pursued by Jon and Robert Kleinberg [**142**]. In this section, we summarize their results.

Fix an integer $d > 0$. We define an on-line random model $\mathcal{PA}_d$ which generates connected graphs $(G_t : t \in \mathbb{N})$, each with constant out-degree d. For simplicity, let $V(G_t) = [t]$. Define G_0 to be isomorphic to K_1. The graph G_{t+1} is defined from G_t by adding a single new vertex $t+1$ that is joined to existing vertices via d randomly chosen edges, sampled independently and with probability

$$\mathbb{P}((s, t+1) \in E(G_{t+1})) = \frac{\deg_{G_t}(s)}{2dt},$$

where $s \in V(G_t)$. This model is similar to the preferential attachment models described in Chapter 4. Note that the graphs generated by $\mathcal{PA}_d$ may have multiple edges, but have no loops. We may orient the graphs G_t in a natural way, with newer vertices always pointing to older ones. Limits of digraphs are defined in a similar way to limits of undirected graphs. Let $\mathcal{PA}_d^\infty$ be the probability space consisting of all the limits of chains of graphs generated by $\mathcal{PA}_d$.

The main questions addressed in [**141**] deal with the isomorphisms types of graphs in $\mathcal{PA}_d^\infty$ for various d. Their main results are summarized in the following theorems.

THEOREM 6.22. *For $d = 1, 2$, with probability 1, $G \in \mathcal{PA}_d^\infty$ is isomorphic (as a directed graph) to a unique digraph R_d^∞.*

For larger $d \geq 2$, the picture radically changes.

THEOREM 6.23. *For $d > 2$, with probability 1, there exist $G, H \in \mathcal{PA}_d^\infty$ which are not isomorphic.*

The contrast between these two results is unexpected, since there is no analogue in the finite models $\mathcal{PA}_d$ of emergent properties differentiating the cases $d < 3$ or $d \geq 3$ (other than the fact that $\mathcal{PA}_d$ generates graphs of arboricity d). The proof of Theorem 6.23 is technical, and relies on inequalities for submartingales. The main idea is to count the number of embeddings of finite graphs. We direct the reader to the paper [**141**] for details. We are content here to sketch a proof of Theorem 6.22. We first mention the following theorem, while omitting the technical proof.

THEOREM 6.24 ([**141**]). *For all vertices $i \in \mathbb{N}$, with probability* 1,

$$\lim_{t\to\infty} \frac{\deg_{G_t}(i)}{\sqrt{t}}$$

exists and is positive.

Consider the unique isomorphism type of tree T which is regular of degree $\aleph_0$. Fix a vertex r which we will call the root. The digraph R_1^∞ is an orientation of T. Label each vertex by its distance to r. The root r has no out-neighbours. Given a vertex u of distance $t > 0$ from r, the unique out-neighbour of u is its unique neighbour on the path connecting u to r. Note that R_1^∞ is the unique isomorphism type of tree with each vertex of infinite in-degree and with the property that each vertex has a directed path to r.

THEOREM 6.25. *With probability* 1, *$G \in \mathcal{PA}_1^\infty$ is isomorphic (when considered as a directed graph) to a unique digraph R_1^∞.*

Proof. By definition, the vertex 0 has out-degree 0, while every other vertex has out-degree 1. If the in-degree of a vertex were finite in G, then

$$\lim_{t\to\infty} \frac{\deg_{G_t}(i)}{\sqrt{t}} = 0$$

contradicting Theorem 6.24. By the definition of $\mathcal{PA}_1^\infty$, each vertex of G has a directed path to 0. □

We now study the case $d = 2$. Consider the following axioms.

(A1) There is a unique vertex u_0 of out-degree 0. Every other vertex has out-degree 2.

(A2) For each pair of (not necessarily distinct) vertices v, w, there are infinitely many vertices $z \neq u_0$ such that (z, v) and (z, w) are directed edges.

(A3) The graph is acyclic and has no infinite one-way directed path.

Given a digraph, we say that an enumeration $\{x_t : t \in \mathbb{N}\}$ of $V(G)$ is *good* if for all $t \in \mathbb{N}$, the out-neighbours of x_t are among $\{x_i : 0 \leq i \leq t-1\}$.

LEMMA 6.26. *If G satisfies (A1) and (A3), then G has a good enumeration.*

Proof. By (A1), we may choose x_0 to be u_0. Assuming

$$\{x_0, x_1, \ldots, x_{t-1}\}$$

were chosen, consider a vertex u of

$$V(G)\backslash\{x_0, x_1, \ldots, x_{t-1}\}.$$

If u has out-degree 0 in $V(G)\backslash\{x_0, x_1, \ldots, x_{t-1}\}$, then let $u = x_t$. Otherwise, by (A3), in

$$V(G)\backslash\{x_0, x_1, \ldots, x_{t-1}\}$$

there is a directed path from u of finite length. The end point v of this path has out-degree 0, and we choose $v = x_t$. □

THEOREM 6.27. *Any two digraphs G and H satisfying (A1), (A2), and (A3) are isomorphic.*

Proof. Let $\{x_t : t \in \mathbb{N}\}$ and $\{y_t : t \in \mathbb{N}\}$ be good enumerations of $V(G)$ and $V(H)$, respectively, which exist by Lemma 6.26. We proceed by a back-and-forth argument, with $f_0(x_0) = y_0$. For $t \geq 0$, suppose that f_t is a partial isomorphism with domain X_t and range Y_t. We will assume that X_t and Y_t are *closed*: all out-neighbours of vertices in the set are in the set itself.

Let x be the lowest indexed vertex of G not in X_t. By hypothesis, both out-neighbours a, b of x are in X_t. By (A2), there is a vertex y whose out-neighbours are $f_t(a)$ and $f_t(b)$ in Y_t. Extend f_t to f_{t+1} by mapping x to y, and letting $X_{t+1} = X_t \cup \{x\}$ and $Y_{t+1} = Y_t \cup \{y\}$. It is straightforward to see that f_{t+1} is an isomorphism, and that the sets X_{t+1} and Y_{t+1} are closed. Then $F = \lim_{t\to\infty} f_t$ is an isomorphism of G with H. □

Following [**141**], we name the unique isomorphism type of digraph in Theorem 6.27 by R_2^∞. A representation of R_2^∞ (akin to R^* for R) is the following. Let X_0 consist of a single vertex x of out-degree 0, along with a countably infinite set of vertices each joined by two parallel edges to x_0 (and not joined to each other). Assume that X_t is defined and countable. Define X_{t+1} so that for all pairs a, b of (not necessarily distinct) vertices from X_t, we add infinitely many vertices z_i, where $i \in \mathbb{N}$, joined to a, b. Let $(R_2^\infty)^* = \lim_{t\to\infty} X_t$.

THEOREM 6.28. *The digraph $(R_2^\infty)^*$ is isomorphic to R_2^∞.*

Proof. Axioms (A1) and (A3) follow from the definition of $(R_2^\infty)^*$. For axiom (A2), consider a pair a, b of vertices of $(R_2^\infty)^*$. Suppose that a, b are in X_t. Then infinitely many vertices with out-neighbours a, b can be found in X_{t+1}. □

The proof of Theorem 6.22 is now obtained by proving that with probability 1, a digraph $G \in \mathcal{PA}_2^\infty$ satisfies (A1), (A2), and (A3). Axioms (A1), (A3) follow from the definition of the model. (A2) is more delicate; see Proposition 4.2 of [**141**].

The isotropic property is defined in the exact same way for limits of digraphs.

COROLLARY 6.29. *The digraph R_2^∞ is isotropic.*

Proof. By Theorem 6.28, we may identify R_2^∞ with the isomorphic representation $(R_2^\infty)^*$. Fix $t_0 > 0$, and let $G = \lim_{t \geq t_0} X_t$. Fix x a vertex in X_{t_0}, and choose infinitely many vertices Y of X_{t_0+1} such that each $y \in Y$ is joined to x by two parallel edges, and Y is independent. The set Y exists by the definition of $(R_2^\infty)^*$. Define X_0' to be the subdigraph induced by $\{x\} \cup Y$, which is isomorphic to X_0. Assume that the set X_t' was defined in G, contains X_0', is isomorphic to X_t, and has age t_0+t+1. It is clear how to define X_{t+1}': add to X_t' all the infinitely many vertices with age $t_0 + (t+1) + 1$ extending each pair a, b in $V(X_t')$. The limit of X_t' is isomorphic to $(R_2^\infty)^*$ and is an induced subgraph of G. □

6.6. The n-Ordered Graphs and Their Limits

Several other random processes lead to interesting limit graphs. For example, consider the following limit graph. For a fixed integer n, let $R_0 \cong K_n$. For some $t \geq 0$, assume that R_t is a finite graph containing R_0. For each subset S of cardinality n in $V(R_t)$, add a new vertex x_S joined only to the vertices of S. The graph R_t along with the new vertices x_S defines the graph R_{t+1}. Let $R^{(n)} = \lim_{t\to\infty} R_t$. We will call $R^{(n)}$ the *infinite random n-ordered graph*, for reasons that will become clear as we proceed (see Corollary 6.34 and Theorem 6.35).

The limit graph $R^{(n)}$ is motivated by on-line web graph models (such as PA models) where new vertices are joined to exactly n existing ones. In our case, we consider the generic case when *all* n-subsets are extended, resulting in the limit graph $R^{(n)}$. The graphs $R^{(n)}$ were studied first by Bonato, Janssen, and Wang [**49**], and our results in this section follow that work.

We consider various isomorphic ways, both deterministic and random, of representing $R^{(n)}$. We begin by supplying some structural information on the graph $R^{(n)}$ itself.

THEOREM 6.30. *Fix a positive integer n.*

(1) *The graph $R^{(n)}$ is strongly n-e.c., but not $(n+1)$-e.c.*

(2) *If G is a strongly n-e.c. graph, then $R^{(n)}$ is an induced subgraph of G.*

(3) $R^{(n)} \leq R^{(n+1)}$.

(4) $\lim_{n\to\infty} R^{(n)} \cong R$.

Proof. For (1), fix an n-subset A and a finite set B in $V(R^{(n)})$, with A and B disjoint. Let $t_0 = \text{age}(A \cup B)$. Then $z_A \in R_{t_0+1}$ is correctly joined to A and B.

Let $S_1, S_2, \ldots, S_{n+1} \subseteq V(R_t)$ be $n+1$ disjoint n-subsets, and, for each S_i, $1 \leq i \leq n+1$, let z_{S_i} be the new vertex in R_{t+1} that extends S_i. Define $S = \{z_{S_i} : 1 \leq i \leq n+1\}$. No vertex of R_{t+1} is joined to all of S. Assume that no vertex of R_j is joined to all of S, where $j \geq t+1$ is fixed. No vertex of R_{j+1} is joined to all of S. Hence, $R^{(n)}$ is not $(n+1)$-e.c.

For item (2), we proceed by induction on t to embed R_t in G. For the inductive step, let R'_t be the copy of R_t in G. For $k \leq n$, by the strongly n-e.c. property for G, for each k-subset S of $V(R'_t)$, there is a vertex z_S of $V(G)\backslash V(R'_t)$ joined to S and to no vertex of $V(R'_t)\backslash S$. Add each of these vertices z_S to $V(R'_t)$ successively, in such a way that they are all pairwise non-joined, and let R'_{t+1} be the resulting subgraph. Then clearly, $R'_{t+1} \cong R_{t+1}$ and $R'_t \leq R'_{t+1}$. Hence, the chain $(R'_t : t \in \mathbb{N})$ has a limit isomorphic to $R^{(n)}$, which in turn is isomorphic to an induced subgraph of G.

Item (3) follows from (2), since $R^{(n+1)}$ is strongly n-e.c. For item (4), first note that the limit is defined with respect to the chain

$$(R^{(n)} : n \in \mathbb{N}\backslash\{0\}),$$

where $R^{(n)}$ is an induced subgraph of $R^{(n+1)}$ (as in (3)). Let

$$G = \lim_{n\to\infty} R^{(n)}.$$

It is sufficient to prove that G is e.c. For this, let A, B be disjoint finite subsets of G. Suppose that $m_1 = |A \cup B|$. Further, suppose that m_2 is such that $(A \cup B) \subseteq V(R^{(m_2)})$. Let $m = m_1 + m_2$. Then $(A \cup B) \subseteq V(R^{(m)})$, and $R^{(m)}$ is m-e.c. Therefore, there is a vertex correctly joined to A, B in $R^{(m)}$, hence in G. □

Let G and H be countable graphs such that $H \leq G$ and $|V(H)| \geq n$. We say that $H \preccurlyeq_{(n,v)} G$ if there is a vertex $v \in V(G)$ such that $H = G - v$ and $\deg_G(v) = n$. Put another way, we delete a vertex of degree n in G to obtain H. We write $H \preccurlyeq_{(n)} G$ if there is countable chain $(G_t : t \in I)$ so that $G_0 \cong H$ and $G_t \preccurlyeq_{(n,v)} G_{t+1}$ for some $v \in G_{t+1}$, and if I is finite, then G equals the G_t with maximum index; if I is infinite, then $G = \lim_{t\to\infty} G_t$.

For example, $K_n \preccurlyeq_{(n)} R^{(n)}$ for all positive integers n. For another illustrative example, consider the unique isomorphism type T of a countable tree such that each vertex has infinite degree. You should check that $K_1 \preccurlyeq_{(1)} T$, T is strongly 1-e.c., and $R^{(1)}$ is isomorphic to T.

Observe that $\preccurlyeq_{(n)}$ is an order relation (that is, a reflexive, antisymmetric, and transitive binary relation) on finite graphs. It is an exercise to show that this also holds for countable graphs (see Exercise 19; the hard part is to show that the relation is transitive). Our next result uses the strongly n-e.c. properties and the relations $\preccurlyeq_{(n)}$ to characterize the isomorphism type of $R^{(n)}$.

THEOREM 6.31. *Let n be a fixed positive integer and let G be a countable graph. Then $G \cong R^{(n)}$ if and only if G is strongly n-e.c. and $K_n \preccurlyeq_{(n)} G$.*

Proof. As the forward direction is immediate, we prove the reverse direction. Suppose, without loss of generality, that $G = \lim_{n\to\infty} G_t$, where $G_0 \cong K_n$, and $G_t \preccurlyeq_{(n)} G_{t+1}$ for all $t \in N$. Enumerate $V(G)\backslash V(G_0)$ as $\{v_t : t \in \mathbb{N}\backslash\{0\}\}$ so that v_t is the unique vertex of G_t not in G_{t-1}.

Let $f_0 : G_0 \to R_0$ be any fixed isomorphism. As the induction hypothesis, suppose that for a fixed $t \geq 0$, there is a finite induced subgraph J_t of G containing G_t along with an isomorphism $f_t : J_t \to R_t$ extending f_0.

The graph R_{t+1} is formed from R_t by extending all n-sets of vertices (that is, for each n-set S in $V(R_t)$ add a new vertex joined to S). If we can find analogous vertices in G for all n-subsets of J_t, including the vertex v_{t+1}, then this will define J_{t+1}.

The vertex v_{t+1} extends some fixed n-set S of G_t. List the n-sets of R_t as $(X_i : 1 \leq i \leq r)$, with $X_1 = f_t(S)$. By induction, we may find vertices a_i extending the n-sets $f_t^{-1}(X_i)$, so that none of the a_i are pairwise joined. More explicitly, let $a_1 = v_{t+1}$. Assuming $a_1, \ldots, a_i$ have been chosen such that $\mathrm{age}(a_j) < \mathrm{age}(a_{j+1})$ for all $1 \leq j \leq i$, choose a_{i+1} to be the first v_k extending $f_t^{-1}(X_{i+1})$ so that $\mathrm{age}(v_k) > \mathrm{age}(a_i)$. Note that v_k is not joined to any v_j, where $j < k$, unless $v_j \in f_t^{-1}(X_{i+1})$. (We use here the fact that v_k is joined to exactly n lower age vertices.) The vertex v_k exists by the strongly n-e.c. property for G. Define J_{t+1} to be the subgraph induced by $V(G_t) \cup \{a_i : 1 \leq i \leq r\}$. It is straightforward to see that $f_{t+1} : J_{t+1} \to R_{t+1}$ is an isomorphism.

Define $f : G \to R^{(n)}$ by $f = \bigcup_{t\in N} f_t$. Then f is an isomorphism by construction. □

Beyond the structural characterization given in Theorem 6.31 the graphs $R^{(n)}$ satisfy self-similarity properties.

COROLLARY 6.32. *For each $n \geq 1$, the graph $R^{(n)}$ is isotropic.*

Proof. Fix $t_0 > 0$. It is straightforward to see that at time t_0+n-1 there is an isomorphic copy R_0' of K_n. For $t > 0$ we may now define R_t' by extending all n-sets in R_{t-1}'. Let $H = \lim_{t\to\infty} R_t' \leq \lim_{t\geq t_0} R_t$. Then H is strongly n-e.c., and $K_n \preceq_{(n)} H$. The assertion now follows from Theorem 6.31 applied to H. □

Can we realize $R^{(n)}$ as the limit of a random process? We next introduce a new random graph process called *Model n*, and which with high probability will generate $R^{(n)}$. The model has some similarities to the Erdős and Rényi model $G(\mathbb{N}, p)$ for R, but there are important differences. In Model n new vertices are added so that older vertices have a higher probability of acquiring new neighbours than their younger counterparts.

For Model n, the single parameter of the model is $n \in \mathbb{N}\backslash\{0\}$. Start with $G_0 \cong K_n$, with vertices labelled $v_1, \ldots, v_n$. For $t \geq 1$ fixed, assume that G_{t-1} has been defined and there are finitely many vertices in G_{t-1}. At time t, add a new vertex v_{n+t}, and choose a set S of n distinct vertices from $V(G_{t-1})$, where the probability that a vertex v_i is included in the set is exponentially proportional to its age. More precisely, for each $S = \{v_{i_1}, \ldots, v_{i_n}\}$, define $\mu(S) = 2^{-(i_1+\cdots+i_n)}$. Define

$$C_t = \sum_{1 \leq j_1 < j_2 < \cdots < j_n \leq t+n-1} 2^{-(j_1+j_2+\cdots+j_n)}.$$

In particular, C_t is the sum of all the $\mu(S)$, where S is a subset of cardinality n from $V(G_{t-1})$. The probability that S is chosen from $V(G_{t-1})$ equals $\mu(S)/C_t$. If S is chosen, then join v_{n+t} to each vertex of S.

THEOREM 6.33. *Let $G = \lim_{t\to\infty} G_t$, where G_t was generated by Model n. With probability* 1, *G is strongly n-e.c.*

Proof. Fix disjoint finite subsets of $V(G)$, A and B, so that $|A| = n$. We prove that the probability that there is no vertex correctly joined to A and B in G is 0. As there are only countably many choices for A and B and a countable union of measure 0 sets is measure 0, the assertion will follow. Let $A = \{v_{i_1}, \ldots, v_{i_n}\}$, so that $\text{age}(v_{i_j}) \leq \text{age}(v_{i_{j+1}})$, for all $1 \leq j \leq n$. Let $t_0 = \text{age}(A \cup B)$. For each $t \geq t_0 + 1$, let V_t be the event that v_{n+t} is joined to all vertices in A and to no vertex in $V(G_{n+t-1})$. Note that the events V_t are pairwise independent. Then

$$\mathbb{P}(V_t) = 2^{-(i_1+\cdots+i_n)}/C_t.$$

Note that $C_t \leq \left(\sum_{j=1}^{t+n-1} 2^{-j}\right)^n \leq 1$ for all t. Therefore, if we let t' be the minimum age of a vertex in $A \cup B$, then

$$\mathbb{P}(V_t) \geq 2^{-(i_1+\cdots+i_n)} \geq 2^{-nt'} \tag{6.2}$$

for all $t \geq t_0$.

Therefore, by (6.2) the probability that there exists no vertex in G that is correctly joined to A and B is at most

$$\mathbb{P}\left(\bigcap_{t=t_0}^{\infty} \overline{V_t}\right) = \prod_{t=t_0}^{\infty}(1 - \mathbb{P}(V_t))$$
$$\leq \lim_{t\to\infty}(1 - 2^{-nt'})^t = 0. \qquad \square$$

In Model n, exactly n edges are added at each time-step from the new vertex to existing vertices. Hence, by Theorems 6.31 and 6.33 we have the following analogue of Theorem 6.1.

COROLLARY 6.34. *With probability* 1, *a limit generated by Model n is isomorphic to $R^{(n)}$.*

Despite these results, many problems remain open about $R^{(n)}$. We say that a graph G *generates* $R^{(n)}$ if the limit graph $R^{(n,G)}$ formed by starting with G and extending all n-sets over time is isomorphic to $R^{(n)}$. By definition, K_n generates $R^{(n)}$. An open problem is to characterize the graphs that generate $R^{(n)}$, even for small graphs. For example, does $\overline{K_2}$ generate $R^{(2)}$?

As is well known, a tree may be defined from K_1 by recursively adding vertices of degree one. A more general notion is an *n-tree*, $n \geq 1$, which is a graph recursively built from K_n by adding a new vertex of degree n joined to an existing n-clique. A *partial n-tree* is a spanning subgraph of an n-tree. Partial n-trees have fascinating connections to treewidth and algorithmic results in graph theory. See [**52**] for more on partial n-trees.

Following [**49**], we say that a countable graph G is *n-ordered* if there exists a well-ordering $(x_i : i \in I)$ of its vertices, where I is finite, or I has the order-type of $\mathbb{N}$ so that each x_j has at most n neighbors x_i, with $i < j$. Expressed differently, a vertex is joined to at most n vertices appearing earlier in the ordering. The ordering $(x_i : i \in I)$ is an *n-ordering* of $V(G)$.

Each finite planar graph is 5-ordered, although K_5 is 5-ordered and not planar. Every partial n-tree is n-ordered, but the converse is false in general for $n \geq 3$. For example, the graph G in Figure 6.6 is 3-ordered, but is not a partial 3-tree. If G is n-ordered, then by the greedy algorithm, $\omega(G), \chi(G) \leq n + 1$.

The n-ordered graphs are essential to understanding the structure of $R^{(n)}$. The following result characterizes the class of isomorphism types of countable induced subgraphs of $R^{(n)}$.

THEOREM 6.35. *A countable graph G is an induced subgraph of $R^{(n)}$ if and only if G is n-ordered.*

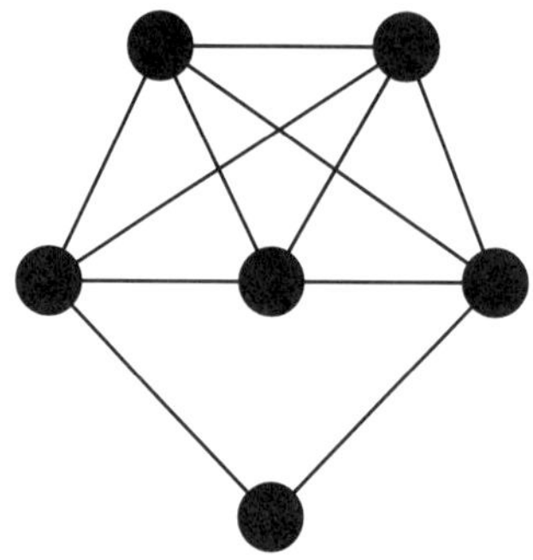

Figure 6.6. The 3-ordered graph G.

Proof. For the forward direction, suppose that $G \leq R^{(n)}$. List the vertices of G according to their age in $R^{(n)}$ from youngest to oldest: $V(G) = (x_i : i \in I)$, where I is finite or $I = \mathbb{N}$. It is not hard to see that this gives an n-ordering of G.

For the reverse direction, let $(x_i : i \in I)$ be a fixed n-ordering of G, so that $G = \lim_{t\to\infty} G_t$, where G_t is the graph induced on $\{x_i : i \leq t\}$. We embed G_t into $R^{(n)}$ inductively. Let G'_0 be the graph induced on a fixed vertex x of $R^{(n)}$. For $t \geq 0$, assume that $G_t \cong G'_t \leq R^{(n)}$, and that G'_t contains x. Let $\text{age}(G'_t) = s$. The vertex x_{t+1} is joined to at most n vertices S in G_t . Let $T = V(G_t) \setminus S$, and let S' and T' be the corresponding subsets of $V(G'_t)$. Let X' be a set in $R^{(n)}$ with cardinality $n - |S'|$ that is disjoint from S' and T'. Suppose that $\text{age}(X') = t'$; without loss of generality, $t' > t$.

Then $z_{S' \cup X'} \in R_{t'+1}$ is joined to S' in R_t and no vertex of T'. Hence,

$$G'_{t+1} = R_{t'+1} \upharpoonright \left(V(G'_t) \cup \{z_{S'}\}\right) \cong G_{t+1},$$

and $G'_t \leq G'_{t+1}$.

It follows that

$$G = \lim_{t\to\infty} G_t \cong \lim_{t\to\infty} G'_t \leq R^{(n)}. \qquad \square$$

COROLLARY 6.36. *The graph $R^{(n)}$ embeds all countable n-ordered graphs.*

Hence, the graph $R^{(n)}$ is a countable universal n-ordered graph, analogous to the infinite random graph, which is a countable universal graph. Furthermore, every strongly n-e.c. graph embeds in $R^{(n)}$ by Theorem 6.30 (2). Like R_H, the graph $R^{(n)}$ is then both a maximal n-ordered graph and a minimal strongly n-e.c. graph (with respect to the embedding order on countable graphs).

The tree $R^{(1)}$ is vertex-transitive (see Exercise 18). However, for $n > 1$, the graph $R^{(n)}$ is not in general vertex-transitive. This is in sharp contrast to R, which possesses a large amount of symmetry. For example, if $n = 2$, then consider the following 2-ordered graph in Figure 6.7. Since every vertex

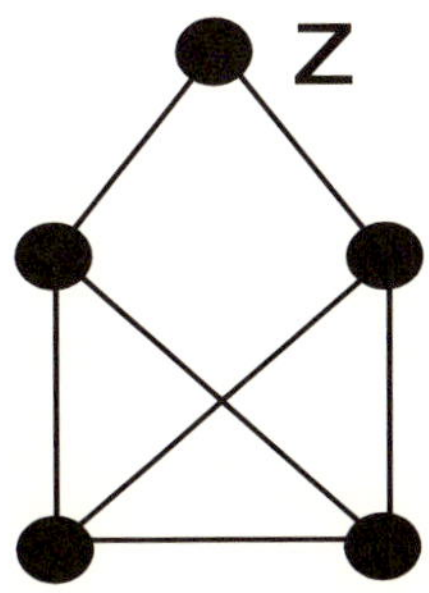

Figure 6.7. The 2-ordered graph G.

but z has degree 3, z may only appear as the last vertex in any 2-ordering of G. If $R^{(2)}$ were vertex-transitive, then we may map z in G to a vertex u of K_2 via an automorphism of G. In particular, there would be an isomorphic copy G' of G in $R^{(2)}$, containing u, and with u acting as z. But then the 2-ordering of $R^{(2)}$ would induce a 2-ordering L of G' with z as the first or second vertex in L, which is a contradiction.

The n-ordered graphs are a new graph class and as such, are of interest in their own right. Given a graph G, a *simple n-reduction* consists of deleting a single vertex of degree at most n. An *n-reduction* consists of a sequence of simple n-reductions. A graph G is an *n-core* if no n-reductions are possible in G. An *n-core of* G is an induced subgraph H such that H is obtained from G by an n-reduction and H is an n-core. See Figure 6.8. If an n-core

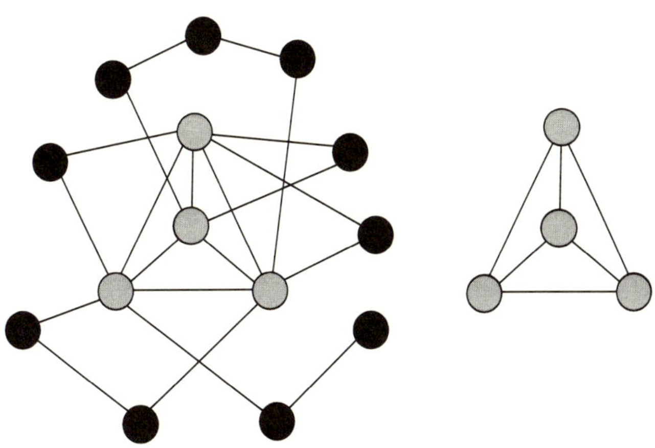

Figure 6.8. A graph and its 2-core.

H of G is non-trivial, then $\delta(H) \geq n+1$. For more on n-cores, see [**34, 182**].

Each finite graph G is $|\triangle(G)|$-ordered. It follows that the graph parameter

$$\Theta(G) = \min\{n : G \text{ is } n\text{-ordered}\}$$

is well defined. We say that $\Theta(G)$ is the *orderability* of G. The $\Theta(G)$-core of G is always K_1, and the location of this K_1 in $V(G)$ need not be unique. It is straightforward to see that

$$\delta(G) \leq \Theta(G) \leq \triangle(G).$$

The following lemma has a straightforward proof.

LEMMA 6.37. *For a finite graph G, and for all non-negative integers n, any two n-cores of G are isomorphic.*

Proof. Without loss of generality, we may assume that $n < \Theta(G)$. Let A and B be n-cores of G so that $V(A) \neq V(B)$. Consider a sequence of n-reductions from G to A, and let x be the first vertex of $V(B) \setminus V(A)$ in the sequence. As no vertex of B has been deleted and $\delta(B) \geq n+1$, the vertex x is joined to $n+1$ vertices yet to be deleted. It is impossible to form an n-reduction by deleting x, which gives a contradiction. Hence, $B \subseteq A$. By symmetry, $A = B$. □

The n-core of G and the orderability of G may be computed in polynomial time. The algorithm for computing the n-core is simple: iteratively delete vertices of degree at most n (in any order). The algorithm for computing $\Theta(G)$ is also straightforward: find the $r_0 = \delta(G)$-core of G; call this G_1. Then find the $r_1 = \delta(G_1)$-core of G_1; iterate this process until K_1 is obtained. If the algorithm terminates in say k steps, then it is easy to see that $\Theta(G) = \max\{r_0, r_1, \ldots, r_{k-1}\}$.

To close the chapter, we introduce a new combinatorial game played on a graph G called *n-deletion*. The game n-deletion is inspired by the game of Cops and Robber discussed at the end of Chapter 7. There are two players: a *deleter* and a *mover*. They move on alternate time-steps, with the deleter beginning the first round of play. The deleter's move consists of a simple n-reduction on G. The mover starts at any vertex of G, and a move for him consists of moving to an adjacent vertex. The mover can never remain on a vertex for more than one round or the deleter wins. In particular, if the mover's position is on an isolated vertex, then the mover loses. If the mover can move indefinitely, even just back-and-forth on an edge, then the deleter loses. A *winning strategy for the mover* is a way of playing where the moves of the mover are determined by the previous moves of both players, and allows the mover to win no matter what the deleter does. Take a moment to convince yourself that there is no strategy possible for the deleter.

THEOREM 6.38. *Let G be a finite graph, and let $n \geq 1$ be fixed. The following are equivalent.*

(1) *The graph G is n-ordered.*
(2) *The n-core of G is K_1.*
(3) *The mover has a winning strategy for n-deletion played on G.*
(4) *There is an acyclic orientation of G so that each vertex has out-degree at most n.*
(5) *The graph G is an induced subgraph of $R^{(n)}$.*

Proof. It is straightforward to see that items (1) and (2) are equivalent. That (1) and (5) are equivalent follows from Theorem 6.35. To see that (1) implies (4), let $(x_i : 1 \leq i \leq r)$ be an n-ordering of G. Orient the vertices so that (x_j, x_i) is a directed edge whenever $i < j$ and $x_i x_j \in E(G)$. Then each vertex has at most n out-neighbours. Since vertices may only point to vertices with smaller index, there are no directed cycles. To prove that (4) implies (1), embed the given acyclic orientation of G into a linear order. The latter ordering is an n-ordering.

We now prove that items (1) and (3) are equivalent. Suppose first that G is n-ordered. The deleter deletes vertices of degree at most n until the resulting graph is K_1. This is possible since G is n-ordered. Either the mover occupies an isolated vertex after one of the deleter's moves, or the mover resides on the K_1 after the deleter's last move. In either case, the deleter wins. Hence, (1) implies (3).

To prove that (3) implies (1), suppose the graph G is not n-ordered. In particular, since (1) and (2) are equivalent, the n-core H of G has more than one vertex. The mover's strategy is to always stay in H. No matter what move the deleter makes, the mover is safe: since vertices of the n-core H have degree at least $n+1$ in H, they are never deleted in any move of the deleter. As $n + 1 \geq 2$, the mover may always move in H. □

Exercises

(1) Prove, using the Axiom of Choice, that every infinite connected graph has a spanning tree.

(2) (a) Show that there are $2^{\aleph_0}$ many non-isomorphic graphs that are mutually embeddable with the tree G in Figure 6.1.
 (b) Give examples of trees T_1 and $T_{\aleph_0}$ so that the cardinality of the set of non-isomorphic graphs mutually embeddable with T_n is n. It is an open problem whether there exist graphs G such that the

set of non-isomorphic graphs mutually embeddable with G is n, where $n \geq 2$ is an integer. See [**51**].

(3) (Requires some background in logic and set theory.) Consider a countable model M of ZF set theory. Such a model exists by the downward Löwenheim-Skolem theorem [**133**]. Define a graph $G(M)$ to have vertices the sets of M, with two sets x and y joined if $x \in y$ or $y \in x$. Prove that $G(M)$ is e.c., and hence, is a representation of R.

(4) [**18**] Define a probability space of random graphs as follows. Fix a real number $p \in (0,1)$. The vertices of the graphs are the points of a countable affine geometry on $\mathbb{Q} \times \mathbb{Q}$. Let ℓ_∞ consist of $\mathbb{Q} \cup \{\infty\}$, identified with slopes of points. For $S \subseteq \ell_\infty$, define $G(S)$ to have vertices $\mathbb{Q} \times \mathbb{Q}$ with two vertices x and y joined if the slope of the line xy is in S. Choose the elements of S from ℓ_∞ independently and with probability p. Name the resulting space of graphs $\mathcal{G}(p)$. Prove that with probability 1, $G(S) \in \mathcal{G}(p)$ is isomorphic to R.

(5) To each subset S of $\mathbb{N}$, let χ_S be its characteristic function. Identify χ_S with a countably infinite 0-1 sequence. Define $X(S)$ to be the graph whose vertices are natural numbers, and i and j are joined if $|i-j| \in S$.
   (a) The function χ_S is *universal* if every finite 0-1 sequence occurs as a consecutive subsequence. If we generate a random χ_S by an infinite sequence of independent fair coin tosses, prove that with probability 1, χ_S is universal.
   (b) Prove that $X(S)$ is isomorphic to R if and only if χ_S is universal.

(6) A tournament is *e.c.* if for all finite disjoint sets of vertices A and B (one of which may be empty), there is a vertex z joined to each vertex of A and joined from each vertex of B.
   (a) Prove that any two e.c. tournaments are isomorphic.
   (b) Define a probability space $T(\mathbb{N},p)$ as follows. For each pair (i,j) of integers with $0 \leq i < j$, independently with probability p add the directed edge (i,j); with probability $1-p$, add (j,i). Prove that with probability 1, $G \in T(\mathbb{N},p)$ is an e.c. tournament. The unique isomorphism type is called the *infinite random tournament.*
   (c) Let the vertices of a tournament G be the set of primes $\mathbb{P}_3$ congruent to 3 (mod 4). Two distinct primes p and q in $\mathbb{P}_3$ are joined if p is a square (mod q) or q is a square (mod p). Prove that G is an e.c. tournament.

(7) [**43**] Let ω be the order-type of $\mathbb{N}$.
   (a) Show that the countable linear orders (that is, transitive tournaments) with $(\mathcal{P})$ are isomorphic to the ordinal powers ω^α, where α is a countable ordinal.

(b) Prove that a tournament with $(\mathcal{P})$ that is not a linear order is isomorphic to the infinite random tournament.

(8) (a) Prove that R is homogeneous.
(b) Show that the Henson graphs H_n, with $n \geq 3$, are universal for K_n-free graphs, and are homogeneous.

(9) Characterize the graphs isomorphic to the spanning subgraphs of R.

(10) Give an example of a countable tree that embeds all countable trees.

(11) Without using the Axiom of Choice, prove that the following graphs contain maximal independent sets.
(a) R.
(b) H_n, where $n \geq 3$ is an integer.
(c) R_H, where H is a finite graph.

(12) (a) Show that for $x \in V(R)$ and $e \in V(R)$, the following graphs are isomorphic to R:

$$R - x, \ R - e, \ \overline{R}.$$

(b) Prove that for all $x \in V(R)$, $\langle N(x) \rangle_R$ and $\langle N^c(x) \rangle_R$ are isomorphic to R. It is an open problem to determine if there any other graphs G with the property that for all vertices x, $\langle N(x) \rangle_G$ and $\langle N^c(x) \rangle_G$ are isomorphic to G.

(13) Prove that R has the pigeonhole property.

(14) (a) Prove that $(\mathcal{P})$ implies the isotropic property.
(b) Give examples of $2^{\aleph_0}$ many non-isomorphic isotropic graphs.

(15) Show that $R_{K_2} \ncong R_{C_6}$.

(16) (a) Determine the diameter of the non-trivial connected component of R_{K_2}.
(b) Show that $R^{(1)}$ has infinite diameter, but $R^{(n)}$ has diameter 2 for all $n \geq 2$.

(17) [**142**] Define a digraph G as follows. For a fixed infinite set S, let the vertices of G consist of all finite rooted binary trees whose edges are labelled with elements of S. If T is a tree with more than one vertex, then T has two outgoing edges in G to its left and right subtrees. Prove that $G \cong R_2^{\infty}$.

(18) A *ray* is an infinite one-way path, while a *Hamilton path* in an infinite graph is a spanning ray.
(a) Prove that R has a Hamilton path.
(b) Prove that the non-trivial component of R_H has a Hamilton path.
(c) Prove that each $R^{(n)}$, where $n \geq 2$, has a Hamilton path.

(19) Show that $R^{(1)}$ is vertex-transitive.

(20) [**49**] Prove that the binary relation $\preccurlyeq_{(n)}$ is transitive on countable graphs.

(21) (a) [**49**] The *Cartesian product* $G\Box H$ of graphs G and H has vertices $V(G) \times V(H)$ and edges $(a,b)(c,d)$ if and only if $b = d$ and $ac \in E(G)$ or $a = c$ and $bd \in E(H)$. Prove that for finite graphs G and H,

$$\Theta(G\Box H) = \Theta(G) + \Theta(H).$$

(b) Here is an open-ended problem. Several other graph products exist, such as the lexicographic, strong, and categorical products (see [**135**]). How does the parameter Θ behave with respect to these products?

(22) [**44**] Prove that each countable group admits an isomorphism to a subgroup of $\mathrm{Aut}(R)$. The corresponding statement for $\mathrm{Aut}(R_H)$ or $\mathrm{Aut}(R^{(n)})$ is an open problem.

(23) (Requires some background in topology.)

(a) [**58**] Prove that $\mathrm{Aut}(R)$ embeds in the group of homeomorphisms of $\mathbb{Q}$ (equipped with the usual topology).

(b) A research problem would be to consider whether the analogous statement in (a) holds if $\mathrm{Aut}(R)$ is replaced by $\mathrm{Aut}(R_H)$ or $\mathrm{Aut}(R^{(n)})$.

Chapter 7

New Directions in Internet Mathematics

I am not young enough to know everything.

–Oscar Wilde

Every human activity, good or bad, except mathematics, must come to an end.

–Paul Erdős

7.1. Introduction

In the last few years there was an explosion of mathematical research related to models of W. With interest emanating from mathematicians and scientists in other disciplines (such as computer science, physics, and biology) a corpus of work is now evolving. A new journal, *Internet Mathematics*, was introduced in 2003 on the topic. While much of the earlier mathematical work on W focused on designing models satisfying certain properties from Chapter 2 such as power law degree distributions, new approaches are constantly emerging.

We survey three recent research directions on modelling W in the present chapter. As we provide only a survey, proofs are omitted; the topics presented here are good candidates for additional study such as a project paper. In the first section, we describe work of Mihail and Papadimitriou [**161**] and Chung, Lu, and Vu [**79**] on the eigenvalues of power law graphs, and explore the so-called eigenvalue power law. As we will learn, in random power law graphs, the distribution of eigenvalues of the adjacency matrix also follows a power law. Section 2 discusses the result of Berger et al. [**26**] on the rigorous modelling of viral spread on W. The aim in [**26**] is to use the SIS model from compartmental epidemiology to model virus spread on scale-free networks. We finish with a description of work from Cooper et al. [**82**] and Bonato et al. [**50**] on the domination number in web graph models, and a discussion of problems related to vertex pursuit games in such models.

7.2. Eigenvalues of Power Law Graphs

Spectral methods in graph theory supply powerful techniques for uncovering properties of graphs. Recall from Chapter 5 that given a graph G of order n with $V(G) = [n]$, the *adjacency matrix* $A(G)$ is the real symmetric $n \times n$ matrix whose (i, j) entry is 1 if $ij \in E(G)$, and 0 otherwise. We witnessed first-hand the utility of spectral methods in our discussion of ranking algorithms in Chapter 5. Not only are spectral techniques in graph theory of use for web mining and ranking algorithms, they have important applications to partitioning and clustering in the web. Owing to the prominence of spectral methods in graph theory and its applications, a fundamental question is therefore the following: what are the eigenvalues of W, or more generally, of power law graphs?

Based on empirical data, [**102**] observed that the first 20 or so of the largest eigenvalues of the adjacency matrix of the internet graph follow a power law degree distribution. The exponent of the power law in this case is called the *eigenexponent*. The authors conjectured a power law degree distribution for the eigenvalues of the adjacency matrices of power law graphs. The work of Mihail and Papadimitriou [**161**] proved this so-called *eigenvalue power law*. From [**161**], they proved that a.a.s. graphs in $G(\mathbf{d})$, where $\mathbf{d}$ is a power law degree sequence with exponent $\beta > 3$, follow a power law with eigenexponent $\sqrt{\beta}$. The largest eigenvalues of power law graphs therefore reveal little structure of the graph. As large eigenvalues are correlated with high degree vertices, they suggest that deleting endvertices will make spectral analysis more fruitful for power law graphs.

The results of [**161**] were generalized by Chung, Lu, and Vu [**79**] to the case when $\beta > 2.5$. They also characterize the largest eigenvalue for the cases when $\beta > 2$. We state their main results here. Given a degree sequence $\mathbf{d} = (d_1, \ldots, d_n)$, define the *expected average degree* d of a graph G in $G(\mathbf{d})$ by

$$d = \frac{1}{n} \sum_{x \in V(G)} \deg(x).$$

For k a positive integer, define

$$\mathrm{Vol}_k(G) = \sum_{x \in V(G)} (\deg(x))^k,$$

and define the *second-order average degree*, written $\widetilde{d}$, by

$$\widetilde{d} = \frac{\mathrm{Vol}_2(G)}{\mathrm{Vol}_1(G)}.$$

THEOREM 7.1 ([**79**]). *For $G \in G(\mathbf{d})$, where $\mathbf{d}$ is a power law degree sequence of length n with exponent β, the following results hold a.a.s.*

(1) *For $\beta \geq 3$, if*

$$m = \Delta(G) > d^2 \log^3 n, \tag{7.1}$$

then the largest eigenvalue of G is $(1 + o(1))\sqrt{m}$.

(2) *For $2.5 < \beta < 3$, if*

$$m > d^{\frac{\beta-2}{\beta-2.5}} \log^{\frac{3}{\beta-2.5}} n, \tag{7.2}$$

then the largest eigenvalue of G is $(1 + o(1))\sqrt{m}$.

(3) *For $2 < \beta < 2.5$, if*

$$m > \log^{\frac{3}{2.5-\beta}} n,$$

then the largest eigenvalue of G is $(1 + o(1))\widetilde{d}$.

(4) *For $\beta > 2.5$,*

$$k < n\left(\frac{d}{m \log n}\right)^{\beta-1},$$

and m sufficiently large satisfying (7.1) and (7.2), the k largest eigenvalues of G satisfy a power law degree distribution with eigen-exponent $2\beta - 1$.

We note that in contrast with Theorem 7.1, Füredi and Komlós [**113**] proved that for $p \in (0, 1)$, the largest eigenvalue of the adjacency matrix of $G(n, p)$ is asymptotically normal, and has eigenvalues satisfying the semi-circle law. Another central approach to spectral techniques in graph theory is via the Laplacian matrix. The Laplacian matrix L of G is defined by

$$L_{i,j} = \begin{cases} \deg(i) & \text{if } i = j, \\ -1 & \text{if } ij \in E(G), \\ 0 & \text{else.} \end{cases}$$

For example, the Laplacian of C_5 equals

$$\begin{pmatrix} 2 & -1 & 0 & 0 & -1 \\ -1 & 2 & -1 & 0 & 0 \\ 0 & -1 & 2 & -1 & 0 \\ 0 & 0 & -1 & 2 & -1 \\ -1 & 0 & 0 & -1 & 2 \end{pmatrix}.$$

The Laplacian is a non-invertible, positive-semidefinite matrix (that is, for all $\mathbf{x} \in \mathbb{R}^{|V(G)|}$, $\mathbf{x}^T L\mathbf{x} > 0$), and so has eigenvalues all non-negative and real. The eigenvalues of the Laplacian of random power law graphs $G(\mathbf{d})$ follow the semicircle law [**79**].

7.3. Modelling Viruses on the Web

A side effect of computing in a small world web is vulnerability to viruses or other malicious software. A *computer virus* may be broadly defined as a program that can replicate itself while executing instructions on infected computers, without the user's knowledge. Modelling computer viruses may lead to new methods to combat them. How can we model computer viruses loose on scale-free networks?

An approach to modelling viruses on the web and internet was given by Berger, Borgs, Chayes, and Saberi [**26**]. The authors presume that a virus is roaming over computers linked via the internet, while anti-virus software is continuously battling the outbreak. Computers with the anti-virus software are not permanently immune to infection; hence, a computer can be reinfected with the same virus until the next sweep of the anti-virus software. They also envision the scenario where viruses are propagating in the presence of sweeps from anti-virus software, but are constantly mutating. Computers that were cured of a particular viral infection may become reinfected by a mutated version of the same virus.

One model of infection in a population of individuals is called the *contact process* or *Susceptible-Infected-Susceptible* (*SIS*) model. Individuals may be healthy (but susceptible) or infected. In the SIS model of virus spread, infected individuals (which could be computers, or people, or animals in some population) return to the susceptible class on recovery because the disease confers no immunity against reinfection. The virus is spread from an infected vertex to its neighbours. Vertices become healthy with rate 1 regardless of how many of their neighbours are infected, but healthy vertices become infected at a rate of a parameter $\lambda > 0$ times the number of infected neighbours. Hence, having many infected neighbours increases the probability a vertex will be infected. See Liggett [**153**] for additional background on the SIS model.

More formally, we may describe the process as a continuous-time Markov chain. A *continuous-time Markov chain* M consists of a sequence of random variables $(X_t : t \in [0, \infty))$ taking values in a finite space S with the property that

$$\mathbb{P}(X_{t_n} = j | X_{t_1} = i_1, \ldots, X_{t_{n-1}} = i_{n-1}) = \mathbb{P}(X_{t_n} = j | X_{t_{n-1}} = i_{n-1}),$$

for all $j, i_1, \ldots, i_{n-1} \in S$ and for any sequence $t_1 < \cdots < t_n$. Note that unlike discrete-time Markov chains, there are uncountably many random variables in the continuous-time case. Continuous-time Markov chains are determined by their *transition rates* q_{ij}, which may be interpreted as the rate at which the process makes a transition to state j when it is in state i. It can be shown that continuous-time Markov chains make transitions between states independently of the past according to a discrete-time Markov chain. However, once the chain enters a state, it remains there, independently of the past, for an exponentially distributed amount of time. See [**122**] for additional background on continuous-time Markov chains.

The *contact process with parameter* $\lambda > 0$ on G is a continuous-time Markov process $(X_t : t \in [0, \infty))$ with state space $V(G)$ that is identified at time t with a subset $A = \{v \in V(G) : X_t(v) = 1\}$. The parameter λ is the *propagation ratio* of the virus. The vertices of A are said to be *infected*, while vertices of $V(G) \backslash A$ are *healthy*. The transition rates for η_t are defined so that a vertex v of A becomes healthy at rate 1, while a vertex u not in A becomes infected with rate $\lambda \deg_H(u)$, where H is the subgraph of G induced by $\{u\} \cup A$. At time $t = 0$, one vertex, called the *root*, is infected. Note that the larger the value of λ or $\deg_H(u)$, the more vertices become infected.

In a finite graph, it can be verified that the set A will eventually become empty with probability 1; that is, the disease will *die out*. A disease becomes an *epidemic* if the time it takes to die out is exponential in the number of vertices.

The use of SIS as a model for virus spreading on the internet and W was first considered by Pastor-Satorras and Vespignani [**172**]. They considered viral spread on a PA model, using simulation and mean-field methods. The authors suggested that in power law graphs with exponent $\beta \leq 3$, no matter how small the parameter λ is, there is a positive probability of having an epidemic (we may say that the *epidemic threshold* is 0).

The work of Berger et al. [**26**] is the first rigorous analysis of a contact process on graphs generated by web graph models. The authors consider a generalization of the LCD PA model described in Chapter 4. Every vertex of graphs in the model may be viewed as a computer linked on the internet, although the vertices may represent other power law networks. The main result of [**26**] is that even if λ is small, there is a positive chance of having an epidemic on scale-free graphs. This explains to some extent why viruses seem to spread so easily on scale-free networks like the web, and corroborates the findings of [**172**].

THEOREM 7.2 ([**26**]). *For every real* $\lambda \in (0,1]$, *there is a positive integer* N *so that for a graph of order greater than* N *generated by the LCD preferential attachment, if* v *is a vertex chosen u.a.r., then the following hold.*

(1) *With probability* $1 - O(\lambda^2)$, v *is such that an infection with root* v *will survive with probability*

$$\lambda^{\Theta\left(\frac{\log(1/\lambda)}{\log\log(1/\lambda)}\right)}.$$

(2) *The probability that the infection with root* v *will survive is*

$$\lambda^{\Theta(1)}.$$

For the proof of Theorem 7.2, which uses a Pólya urn representation of graphs in the LCD model, the reader is directed to [**26**].

7.4. Dominating Sets in the Web Graph

Domination is an extensively studied topic within graph theory and its applications, and is the focus of hundreds of papers and books (see the book [**128**], which is a good reference for the topic). Recall that a set of vertices S is a *dominating set* in G if each vertex not in S is joined to some vertex of S. The *domination number* of G, written $\gamma(G)$, is the minimum cardinality of a dominating set in G.

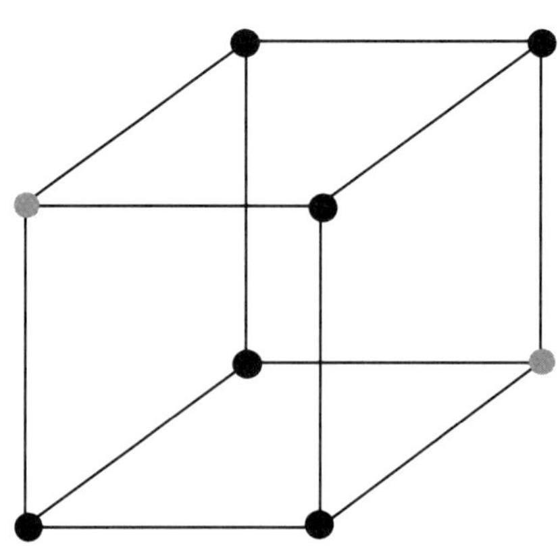

Figure 7.1. $\gamma(Q_3) = 2$. Each pair of antipodal vertices is a dominating set in the cube.

Applications of domination play a role in distributed computing and ad hoc networks (see [**14, 187, 91**]). We can view the dominating set S as vertices providing some service to the whole network. To minimize the cost of the service, $|S|$ is minimized. For example, a web crawler may use a dominating set as a means to quickly reach all vertices in W.

The domination number of $G(n,p)$ was first studied in Dreyer's doctoral thesis [**89**]. The following result of [**89**] gives bounds for the domination number of a finite random graph with p a fixed constant.

THEOREM 7.3 ([**89**]). *Let* $0 < p < 1$ *be fixed and* $q = \frac{1}{1-p}$. *For every real* $\varepsilon > 0$, *with probability* 1 *as* $n \to \infty$, *for* $G \in G(n,p)$,

$$(1-\varepsilon)\log_q n \leq \gamma(G) \leq (1+\varepsilon)\log_q n.$$

The following result was proved in [**117**]. We adopt the notation $\mathbb{L}n = \log_{1/(1-p)} n$.

THEOREM 7.4 ([**117**]). *A.a.s.* $G \in G(n,p)$ *has* $\gamma(G)$ *equalling one of*

$$\lfloor \mathbb{L}n - \mathbb{L}((\mathbb{L}n)(\log n)) \rfloor + 1 \quad \textit{or} \quad \lfloor \mathbb{L}n - \mathbb{L}((\mathbb{L}n)(\log n)) \rfloor + 2.$$

Despite the fact that deterministic graphs can have large domination number relative to the number of vertices (for instance, $\gamma(P_n) = \lceil n/3 \rceil$), Theorem 7.4 demonstrates that a.a.s. $G \in G(n,p)$ has domination number equalling $(1+o(1))\mathbb{L}n = \Theta(\log n)$. The domination number for random graphs in the more general setting when p is a function of n was studied in [**117**].

Domination was studied in the context of a preferential attachment model in [**82**]. Algorithms were supplied for dominating sets in the graphs generated by a PA model. A key interpretation of the results of [**82**] for W is that W has large dominating sets, and so crawlers utilizing a dominating set will need a large sample for accuracy.

The authors consider a variant of the usual PA model with parameter a positive integer m. At time 0, there is a single vertex with m loops. At time $t+1$, a new vertex is introduced to G_t which is joined to existing ones via m edges, whose endpoints u are chosen with probability following a typical preferential attachment rule

$$\frac{\deg_{G_t}(u)}{2mt}.$$

We now describe Algorithm 3 from [**82**], which has an integer parameter $k > 0$. The algorithm is on-line, and dominating sets are grown as new vertices are introduced. At the beginning of the algorithm, let $S = \{v_0\}$. When v_t is born, let Z be the set of all neighbours of v_t in $V(G_t)\backslash S$ of degree $km+1$. If Z is non-empty, then add all vertices of Z to S. Otherwise, if v_t

is not dominated by some element of S, then a vertex of maximum degree in $N_{G_t}(v_t)$ is added to S. It is clear by the way that S is defined that it is a dominating set of G_t.

Let X_t be the random variable equalling the cardinality of the dominating set S generated by Algorithm 3 when run on the aforementioned PA model until time t.

THEOREM 7.5 ([**82**]). *For the PA model, $X_t \sim xt$, where x is a real constant depending on k and m.*

Hence, there is a dominating set which is of cardinality linear in the number of vertices of G_t. There is a similar lower bound which is a linear function of t.

THEOREM 7.6 ([**82**]). *There is a function $f(x, y) : \mathbb{R} \to \mathbb{R}$ such that for every positive integer m, there is a $d(m) > 0$ defined by $f(m, d(m)) = 1$, such that a.a.s. there is no dominating set in the graph G_t of cardinality at most dt.*

Theorems 7.5 and 7.6 demonstrate that a.a.s. $\gamma(G_t) = \Theta(t)$. The constants d and x are usually distinct, but are close as described for some small values of m. For example, when $m = 2$, the authors report that $d = 0.10098$, while the corresponding value of x from Theorem 7.5 (for all values of $k \leq 2$) is 0.2342.

In power law graphs, there may exist an abundance of isolated vertices, even as many as $\Theta(n)$. Since the domination number is bounded from below by the number of isolated vertices, we expect the domination number of a typical power law graph to be around cn, for a constant $c \in (0, 1)$. Recent work from [**50**] shows that this is indeed the case for the off-line random power law graphs $G(\mathbf{d})$, which we now recall from Chapter 4.

Given $\beta > 2$, $d > 0$, and a function $M = M(n)$ (with M tending to infinity with n), we consider the random graph with given expected degrees $d_i > 0$, where

$$d_i = ci^{-\frac{1}{\beta-1}} \tag{7.3}$$

for i satisfying $i_0 \leq i < n + i_0$. The term c depends on β and d, and i_0 depends also on M; namely,

$$c = \left(\frac{\beta-2}{\beta-1}\right) dn^{\frac{1}{\beta-1}}, \qquad i_0 = n\left(\frac{d}{M}\left(\frac{\beta-2}{\beta-1}\right)\right)^{\beta-1}. \tag{7.4}$$

It is not hard to show (see [**76, 78**]) that a.a.s. the random graphs with the expected degrees satisfying (7.3) and (7.4) follow a power law degree distribution with exponent β, average degree $(1 + o(1))d$, and maximum degree $(1 + o(1))M$.

As the number of isolated vertices is a lower bound for the domination number, the following result of [**50**] proves that the domination number of $G(\mathbf{d})$ is a.a.s. equal to $\Theta(n)$.

THEOREM 7.7 ([**50**]). *For a random power law graph $G(\mathbf{d})$ with exponent $\beta > 2$ and average degree d, for all $\varepsilon > 0$, a.a.s. the following hold.*

(1) *If X is the random variable denoting the number of isolated vertices in $G(\mathbf{d})$, then*

$$X = (1+o(1))n \int_0^1 \exp\left(-d\frac{\beta-2}{\beta-1}x^{-1/(\beta-1)}\right) dx.$$

(2) *For $a \in (0,1)$, define*

$$f(a) = a + \int_a^1 \exp\left(-d\frac{\beta-2}{\beta-1}a^{(\beta-2)/(\beta-1)}x^{-1/(\beta-1)}\right) dx.$$

Then

$$\gamma(G) \leq (1+o(1))n \min_{0<a<1} f(a).$$

The theorem demonstrates that the domination number of random power law graphs is of much larger order than the logarithmic domination number of $G(n,p)$ random graphs as described in Theorems 7.3 and 7.4. Hence, we should expect that in real-world power law graphs such as the web graph the domination number is large relative to the number of vertices, and it would be interesting to conduct experiments which corroborate this claim.

The integrals in the statement of Theorem 7.7 do not possess closed-form solutions in general. For example, when $d = 10$ and $\beta = 2.1$ (the observed power law exponent of W; see Chapter 2), the lower and upper bounds are 0.1806 and 0.2940, respectively.

A compelling graph parameter related to the domination number is the so-called cop (or search) number of a graph. The game of *Cops and Robber* is a vertex pursuit game played on a graph G. There are two players, a set of k *cops* (or *searchers*) $\mathcal{C}$, where $k > 0$ is a fixed integer, and the *robber* $\mathcal{R}$. The cops begin the game by occupying a set of k vertices, and the cops and robber move in alternate *rounds*. More than one cop is allowed to occupy a vertex, and the players may *pass*, that is, remain on their current vertex. The players know each other's current locations and can remember all the previous moves; that is, the game is played with *perfect information*. The cops win and the game ends if at least one of the cops can eventually occupy the same vertex as the robber; otherwise, $\mathcal{R}$ wins.

As placing a cop on each vertex guarantees that the cops win, we may define the *cop number*, written $c(G)$, which is the minimum number of cops needed to win on G. The cop number was introduced by Aigner and Fromme [**7**] who proved that if G is planar, then $c(G) \leq 3$.

Whether $c(G) \leq k$, for k a fixed integer, is polynomial time computable (as a function of the number of vertices); see [**24, 119, 126**]. So-called *cop-win* graphs (that is, graphs G with $c(G) = 1$) were structurally characterized in [**169, 170**]. See Figure 7.2 for a cop-win graph. If x is a vertex, then define $N[x]$ (the *closed neighbourhood of* x) to be x along with the vertices joined to x. The cop-win graphs are exactly those graphs which are *dismantlable*: there exists a linear ordering $(x_j : 1 \leq j \leq n)$ of the vertices so that for all $2 \leq j \leq n$, there is an $i < j$ such that $N[x_j] \subseteq N[x_i]$. No analogous characterization (using only local structural information) of graphs with cop number k, where $k > 1$ is a fixed integer, is known; this is a central open problem in the subject. For a survey of results on the cop number and related search parameters for graphs, see [**13**].

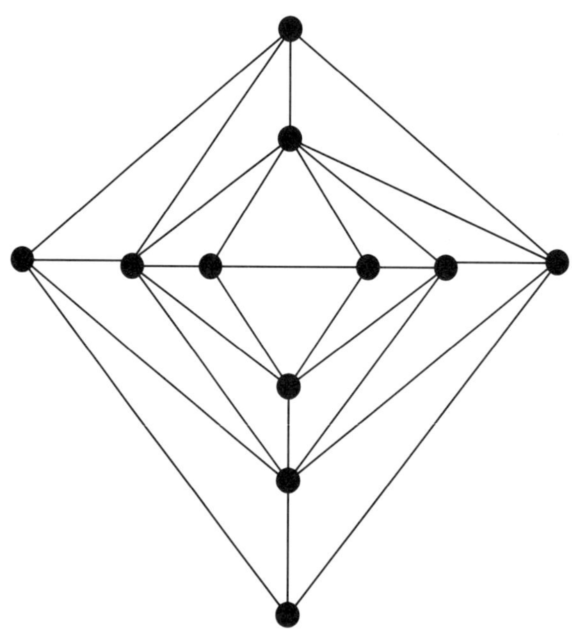

Figure 7.2. A cop-win graph.

Cops and Robber may be thought of as a simplified model for network security. For example, we may think of the robber as an unwanted intruder (such as a virus or hacker) on some real-world network like W. The intruder is moving from vertex to vertex. The goal is then to minimize the resources (that is, number of cops) required to capture the intruder.

Since

$$c(G) \leq \gamma(G) \tag{7.5}$$

(place a cop on each vertex of dominating set with minimum cardinality), the results of this section give an upper bound to the cop number. However, if $n \geq 1$, then $c(P_n) = 1$ and $\gamma(P_n) = \left\lceil \frac{n}{3} \right\rceil$. The bound of (7.5), while useful, is far from tight in general.

The cop number of $G(n,p)$ was studied in [**45**], where the following result was proved.

THEOREM 7.8. *Let $0 < p < 1$ be fixed. For every real $\varepsilon > 0$, with probability 1 as $n \to \infty$, for $G \in G(n,p)$,*

$$(1-\varepsilon)\log_{\frac{1}{1-p}} n \leq c(G) \leq (1+\varepsilon)\log_{\frac{1}{1-p}} n.$$

The upper bound follows from Theorem 7.3. The proof of Theorem 7.8 is obtained by establishing the lower bound for the cop number of $G(n,p)$. For a positive integer k, a graph is $(1,k)$-*e.c.* if for each k-element subset S of vertices of G and vertex u, there is a vertex $z \notin S$ not joined to any vertex in S and joined to u. This is a generalization of the 1-e.c. property described in Chapter 6. We need the following lemma.

LEMMA 7.9. *Let $0 < p < 1$ and $r > 0$ be fixed. For any fixed $0 < \epsilon < 1$, if $\left(\log \frac{1}{1-p}\right) d = 1 - \epsilon$, then*

$$\lim_{n\to\infty} n^{\lfloor d\log n\rfloor+1}\left(1 - r(1-p)^{\lfloor d\log n\rfloor}\right)^{n-\lfloor d\log n\rfloor-1} = 0. \tag{7.6}$$

Proof. It is enough to prove that

$$\lim_{n\to\infty} n^{d\log n+1}\left(1 - r(1-p)^{d\log n}\right)^{n-d\log n-1} = 0. \tag{7.7}$$

If $q = 1-p$ and $c = \log(\frac{1}{q})$, then $c, d > 0$ and $0 < cd < 1$. Now

$$\begin{aligned} &n^{d\log n+1}(1-r(1-p)^{d\log n})^{n-d\log n-1}\\ &= n^{d\log n+1}(1-rq^{d\log n})^{n-d\log n-1}\\ &= n^{d\log n+1}\left(1-\frac{r}{n^{cd}}\right)^{n-d\log n-1}\\ &= \exp\left(f(n)\right), \end{aligned} \tag{7.8}$$

where

$$f(n) = (d\log n + 1)\log(n) + (n - d\log n - 1)\log\left(1 - \frac{r}{n^{cd}}\right).$$

For a sufficiently large n, $\log\left(1-\frac{r}{n^{cd}}\right) < 0$, we have that

$$\lim_{n\to\infty} f(n) = -\infty. \tag{7.9}$$

By (7.8) and (7.9), we have (7.7). □

Proof of Theorem 7.8. Convince yourself that if G is $(1,k)$-e.c., then $c(G) \geq k$. Let $0 < \epsilon < 1$ be fixed, and let $k = \lfloor (1-\epsilon)\log_q n \rfloor$, where $q = \frac{1}{1-p}$. The probability that G is not $(1,k)$-e.c. is at most

$$f(n) = n^{k+1}(1-p(1-p)^k)^{n-k-1}.$$

Setting $r = p$ in (7.6), we have that

$$\lim_{n\to\infty} f(n) = 0$$

by Lemma 7.9. Hence, with probability 1 as $n \to \infty$,

$$c(G) \geq (1-\epsilon)\log_q n.$$ □

What is the cop number of graphs generated by models of W, such as preferential attachment or copying models? This is an inviting open problem, and so serves as an appropriate place to finish.

Exercises

(1) (a) Prove that every rational eigenvalue of a graph is an integer.
 (b) Prove that if the characteristic polynomial of $A(G)$ is irreducible over $\mathbb{Q}$, then the graph G is asymmetric (that is, has no non-trivial automorphisms).

(2) Recall from Chapter 3 that a k-regular graph G with v vertices, such that each pair of joined vertices has exactly λ common neighbours, and each pair of non-joined vertices has exactly μ common neighbours is called a *strongly regular graph*; we say that G is an SRG(v,k,λ,μ).
 (a) If G is an SRG(v,k,λ,μ), then the following two numbers are integers:

$$\frac{1}{2}\left(v-1 \pm \frac{(v-1)(\mu-1)-2k}{\sqrt{(\mu-\lambda)^2+4(k-\mu)}}\right).$$

 Hint: Examine the eigenvalues of G and their algebraic multiplicities.
 (b) [**196**] Prove the *Friendship Theorem*: If G is a graph where each pair of distinct vertices has a unique common neighbour, then G has a universal vertex. **Hint**: Consider the cases when G is regular (apply part (a)) and non-regular (show that two distinct non-joined vertices have the same degree).
 (c) Find an infinite family of graphs satisfying the hypothesis of the Friendship Theorem.

(3) (a) Let q be a prime congruent to 1 modulo 4. Prove that the eigenvalues of the Paley graph P_q are

$$\sum_{k\in\mathbb{Z}_q} \exp\left(\frac{2\pi ijk^2}{q}\right),$$

where $1 \leq j \leq q-1$, and i is the imaginary unit.

(b) Show that the eigenvalues of the Paley graph P_q are all equal to one of

$$\frac{\pm\sqrt{q}-1}{2} \quad \text{or} \quad \frac{q-1}{2}.$$

(4) Given a graph G with vertices $[n]$ without isolated vertices, define the *normalized Laplacian* matrix $\mathcal{L}$ of G by

$$\mathcal{L}_{i,j} = \begin{cases} 1 & \text{if } i=j \text{ and } \deg(i)\neq 0, \\ -\frac{1}{\sqrt{\deg(i)\deg(j)}} & \text{if } ij\in E(G), \\ 0 & \text{else.} \end{cases}$$

(a) Prove that the eigenvalues of $\mathcal{L}$ are in the interval $[0,2]$.

(b) The multiplicity of the eigenvalue 0 is equal to the number of connected components of G.

(5) [**26**] Let G be a star $K_{1,n}$ with center u. If we consider the SIS model described in Section 7.3 with infection parameter λ, then define A_t to be the set of vertices infected at time t. Prove that there exists a C such that if $A_0 = \{u\}$, then for all k

$$\mathbb{P}(A_{\exp(Ck\lambda^2)} \neq \varnothing) = 1 - o(k).$$

(6) (a) Find the domination number of P_n, C_n, Q_n and the Petersen graph.

(b) Prove that if G is connected of order n, then $\gamma(G) \leq \frac{n}{2}$.

(c) Show that for a graph G of order n,

$$\frac{n}{1+\Delta(G)} \leq \gamma(G) \leq n - \Delta(G).$$

(7) (a) Recall that the *Cartesian product* $G\square H$ of graphs G and H has vertices $V(G)\times V(H)$ and edges $(a,b)(c,d)$ if and only if $ac\in E(G)$ or $a=c$ and $bd\in E(H)$. Prove that for graphs G and H,

$$\gamma(G\square H) \leq \min\{\gamma(G)|V(H)|, \gamma(H)|V(G)|\}.$$

(b) An important unsettled conjecture in graph domination is due to Vizing, which states

$$\gamma(G)\gamma(H) \leq \gamma(G\square H).$$

Find an infinite family of graphs $(G_i, H_i : i\in\mathbb{N})$ for which

$$\gamma(G_i\square H_i) = \gamma(G_i)\gamma(H_i).$$

(8) A graph H is a *retract* of G if H is an induced subgraph of G such that there is a homomorphism $f : G \to H$ with the property that $f(u) = u$ for all $u \in V(H)$.

(a) Suppose that H is a retract of G. Show that if the robber has a winning strategy in H, then he has a winning strategy in G.

(b) [**169**] Recall that a graph is *dismantlable* if there exists a linear ordering $(x_j : 1 \leq j \leq n)$ of the vertices so that for all $2 \leq j \leq n$, there is an $i < j$ such that $N[x_j] \subseteq N[x_i]$. Prove that a graph G is dismantlable if and only if G is cop-win.

(c) Verify that the graph of Figure 7.2 is cop-win. What is a winning strategy for the cop?

(9) [**191**] For graphs G and H, prove that

$$c(G \square H) \leq c(G) + c(H).$$

Bibliography

[1] J. Abello, A.L. Buchsbaum, J.R. Westbrook, A functional approach to external graph algorithms, In: *Proceedings of 6th European Symposium on Algorithms*, 1998.

[2] L.A. Adamic, The small world web, In: *Proceedings of the 3rd European Conference on Research and Advanced Technology for Digital Libraries*, 1999.

[3] A. Adler, M. Mitzenmacher, Towards compressing web graphs, In: *Proceedings of the Data Compression Conference*, 2001.

[4] W. Aiello, F. Chung, L. Lu, A random graph model for massive graphs, *Experimental Mathematics* **10** (2001) 53-66.

[5] W. Aiello, F. Chung, L. Lu, Random evolution in massive graphs, *Handbook on Massive Data Sets* (James Abello et al. eds.), Kluwer Academic Publishers, 2002, pp. 97-122.

[6] W. Aiello, A. Bonato, C. Cooper, J. Janssen, P. Prałat, A spatial web graph model with local influence regions, In: *Proceedings of the 5th Workshop on Algorithms and Models for the Web-Graph*, 2007.

[7] M. Aigner, M. Fromme, A game of cops and robbers, *Discrete Applied Mathematics* **8** (1984) 1-12.

[8] R. Albert, H. Jeong, A. Barabási, Diameter of the world-wide web, *Nature* **401** (1999) 130.

[9] D. Alderson, J.C. Doyle, L. Li, W. Willinger, Towards a theory of scale-free graphs: definition, properties, and implications, *Internet Mathematics* **2** (2005) 431-523.

[10] D.J. Aldous, A tractable complex network model based on the stochastic mean-field model of distance, In: *Complex Networks* (E. ben-Naim et al. ed.), Springer Lecture Notes in Physics **650** (2004) 51-87.

[11] N. Alon, M. Krivelevich, The concentration of the chromatic number of random graphs, *Combinatorica* **17** (1997) 303-313.

[12] N. Alon, J. Spencer, *The Probabilistic Method*, Wiley, New York, 2000.

[13] B. Alspach, Sweeping and searching in graphs: a brief survey, *Matematiche* **59** (2006) 5-37.

[14] K. Alzoubi, P.J. Wan, O. Frieder, Message-optimal connected dominating sets in mobile ad hoc networks, In: *Proceedings of the 3rd International Symposium on Mobile ad hoc Networking and Computing*, 2002.

[15] R. Andersen, F.R.K. Chung, L. Lu, Modeling the small-world phenomenon with local network flow, *Internet Mathematics* **2** (2006) 359-385.

[16] K. Azuma, Weighted sums of certain dependent random variables, *Tôhoku Mathematical Journal* **19** (1967) 357-367.

[17] C.A. Baker, A. Bonato, J.M.N. Brown, T. Szőnyi, Graphs with the n-e.c. adjacency property constructed from affine planes, accepted to *Discrete Mathematics*.

[18] C.A. Baker, A. Bonato, N.A. McKay, Graphs with the n-e.c. adjacency property constructed from resolvable designs, submitted.

[19] P. Baldi, P. Frasconi, P. Smyth, *Modeling the Internet and the Web, Probabilistic Methods and Algorithms*, John Wiley & Sons, Ltd, Chichester, West Sussex, England, 2003.

[20] A. Barabási, *Linked: How Everything Is Connected to Everything Else and What It Means*, Perseus Publishing, Cambridge, MA, 2002.

[21] A. Barabási, R. Albert, Emergence of scaling in random networks, *Science* **286** (1999) 509-512.

[22] A. Barabási, E. Ravasz, T. Vicsek, Deterministic scale-free networks, *Physica A* **299** (2001) 559-564.

[23] L. Becchetti, C. Castillo, The distribution of PageRank follows a power law only for particular values of the damping factor, In: *Proceedings of the 15th International Conference on the World Wide Web*, 2006.

[24] A. Berarducci, B. Intrigila, On the cop number of a graph, *Advances in Applied Mathematics* **14** (1993) 389-403.

[25] N. Berger, J. Chayes, C. Borgs, R. D'Souza, R.D. Kleinberg, Competition-induced preferential attachment, In: *Proceedings of the 31st International Colloquium on Automata, Languages and Programming*, 2004.

[26] N. Berger, J. Chayes, C. Borgs, A. Saberi, On the spread of viruses on the Internet, In: *Proceedings of the 16th ACM-SIAM Symposium on Discrete Algorithms*, 2005.

[27] K. Bharat and M. Henzinger, Improved algorithms for topic distillation in a hyperlinked environment, In: *Proceedings of the 21st International ACM-SIGIR Conference on Research and Development in Information Retrieval*, 1998.

[28] D.K. Blandford, G.E. Blelloch, I.A. Kash, Compact representations of separable graphs, In: *Proceedings of ACM-SIAM Symposium on Discrete Algorithms*, 2003.

[29] A. Blass, G. Exoo, F. Harary, Paley graphs satisfy all first-order adjacency axioms, *Journal of Graph Theory* **5** (1981) 435-439.

[30] B. Bollobás, The chromatic number of random graphs, *Combinatorica* **8** (1988) 49-55.

[31] B. Bollobás, *Modern Graph Theory*, Springer-Verlag, New York, 1998.

[32] B. Bollobás, *Random graphs, Second edition,* Cambridge Studies in Advanced Mathematics **73**, Cambridge University Press, Cambridge, 2001.

[33] B. Bollobás, C. Borgs, T. Chayes, O. Riordan, Directed scale-free graphs, In: *Proceedings of the 14th ACM-SIAM Symposium on Discrete Algorithms*, 2003.

[34] B. Bollobás, O. Riordan, J. Spencer, G. Tusnády, The degree sequence of a scale-free random graph process, *Random Structures and Algorithms* **18** (2001) 279-290.

[35] B. Bollobás, O. Riordan, Mathematical results on scale-free graphs, *Handbook of Graphs and Networks* (S. Bornholdt, H. Schuster eds.), Wiley-VCH, Berlin, 2002, pp. 1-32.

[36] B. Bollobás, O. Riordan, Robustness and vulnerability of scale-free random graphs, *Internet Mathematics* **1** (2004) 1-35.

[37] B. Bollobás, O. Riordan, The diameter of a scale-free random graph, *Combinatorica* **24** (2004) 5-34.

[38] B. Bollobás, A. Thomason, Graphs which contain all small graphs, *European Journal of Combinatorics* **2** (1981) 13-15.

[39] B. Bollobás, A. Thomason, Threshold functions, *Combinatorica* **7** (1987) 35-38.

[40] A. Bonato, A survey of models of the web graph, In: *Proceedings of Combinatorial and Algorithmic Aspects of Networking*, 2004.

[41] A. Bonato, K. Cameron, On an adjacency property of almost all graphs, *Discrete Mathematics* **231** (2001) 103-119.

[42] A. Bonato, K. Cameron, On an adjacency property of almost all tournaments, *Discrete Mathematics* **306** (2006) 2327-2335.

[43] A. Bonato, P.J. Cameron, D. Delić, Tournaments and orders with the pigeonhole property, *Canadian Mathematical Bulletin* **43** (2000) 397-405.

[44] A. Bonato, D. Delić, I. Dolinka, All countable monoids embed into the monoid of the infinite random graph, submitted.

[45] A. Bonato, G. Hahn, C. Wang, The cop density of a graph, *Contributions to Discrete Mathematics* **2** (2007) 133-144

[46] A. Bonato, J. Janssen, Infinite limits of copying models of the web graph, *Internet Mathematics* **1** (2004) 193-213.

[47] A. Bonato, J. Janssen, Infinite limits and adjacency properties of a generalized copying model, accepted to *Internet Mathematics*.

[48] A. Bonato, J. Janssen, Limits and power laws of models for the web graph and other networked information spaces, In: *Proceedings of Combinatorial and Algorithmic Aspects of Networking*, 2004.

[49] A. Bonato, J. Janssen, C. Wang, The n-ordered graphs—a new graph class, submitted.

[50] A. Bonato, P. Prałat, C. Wang, Vertex pursuit games in stochastic network models, In: *Proceedings of the Fourth Workshop on Combinatorial and Algorithmic Aspects of Networking*, 2007.

[51] A. Bonato, C. Tardif, Mutually embeddable graphs and the Tree Alternative conjecture, *Journal of Combinatorial Theory, Series B*, **96** (2006) 874-880.

[52] A. Brandstädt, V.B. Le, J.P. Spinrad, *Graph Classes, a Survey*, SIAM Monographs on Discrete Mathematics and Applications, 1999.

[53] S. Brin, L. Page, Anatomy of a large-scale hypertextual web search engine, In: *Proceedings of the 7th International World Wide Web Conference*, 1998.

[54] A. Broder, R. Kumar, F. Maghoul, P. Raghavan, S. Rajagopalan, R. Stata, A. Tomkins, J. Wiener, Graph structure in the web, *Computer Networks* **33** (2000) 309-320.

[55] K. Bryan, T. Leise, The $25,000,000,000 eigenvector: the linear algebra behind Google, *SIAM Review* **48** (2006) 569-581.

[56] P.G. Buckley, D. Osthus, Popularity based random graph models leading to a scale-free degree sequence, *Discrete Mathematics* **282** (2004) 53-68.

[57] L. Caccetta, P. Erdős, K. Vijayan, A property of random graphs, *Ars Combinatoria* **19** (1985) 287-294.

[58] P.J. Cameron, *Oligomorphic permutation groups*, Cambridge University Press, Cambridge, 1990.

[59] P.J. Cameron, *Combinatorics: Topics, Techniques, Algorithms*, Cambridge University Press, Cambridge, 1994.

[60] P.J. Cameron, The random graph, *Algorithms and Combinatorics* (R.L. Graham and J. Nešetřil, eds.), Springer Verlag, New York, 1997, pp. 333-351.

[61] P.J. Cameron, The random graph revisited, *Progress in Mathematics* **201** (2001) 267-274.

[62] P.J. Cameron, The Rado graph and the Urysohn space, submitted.

[63] P.J. Cameron, D. Stark, A prolific construction of strongly regular graphs with the n-e.c. property, *Electron. J. Combin.* **9** (2002) no. 1, Research Paper 31, 12 pp.

[64] G. Cantor, Beiträge zur Begründung der transfiniten Mengenlehre, *Mathematische Annalen* **49** (1897) 207-246.

[65] S. Chakrabarti, *Mining the Web, Discovering Knowledge from Hypertext Data*, Morgan Kauffman Publishers, San Francisco, 2003.

[66] S. Chakrabarti, A. Frieze, J. Vera, The influence of search engines on preferential attachment, In: *Proceedings of the Sixteenth Annual ACM-SIAM Symposium on Discrete Algorithms*, 2005.

[67] F.R.K. Chung, Quasi-random classes of hypergraphs, *Random Structures Algorithms* **1** (1990) 363-382.

[68] F.R.K. Chung, Constructing random-like graphs, *Probabilistic Combinatorics and Its Applications* (B. Bollobás ed.), American Mathematical Society, Providence, 1991, pp. 21-55.

[69] F.R.K. Chung, G. Dewey, D.J. Galas, L. Lu, Duplication models for biological networks, *Journal of Computational Biology* **10** (2003) 677-688.

[70] F.R.K. Chung, R.L. Graham, R.W. Wilson, Quasi-random graphs, *Combinatorica* **9** (1989) 345-362.

[71] F.R.K. Chung, R.L. Graham, Quasi-random hypergraphs, *Random Structures and Algorithms* **1** (1990) 105-124.

[72] F.R.K. Chung, R.L. Graham, Quasi-random tournaments, *Journal of Graph Theory* **15** (1991) 173-198.

[73] F.R.K. Chung, R.L. Graham, Sparse quasi-random graphs, *Combinatorica* **22** (2002) 217-244.

[74] F.R.K. Chung, R.L. Graham, Quasi-random graphs with given degree sequences, accepted to *Random Structures and Algorithms.*

[75] F.R.K. Chung, L. Lu, Connected components in random graphs with given degree sequences, *Annals of Combinatorics* **6** (2002) 125-145.

[76] F.R.K. Chung, L. Lu, The average distances in random graphs with given expected degrees, *Internet Mathematics* **1** (2003) 91-113.

[77] F.R.K. Chung, L. Lu, Coupling on-line and on-line analyses for random power law graphs, *Internet Mathematics* **1** (2004) 409-461.

[78] F.R.K. Chung, L. Lu, *Complex graphs and networks*, American Mathematical Society, Providence, RI, 2006.

[79] F.R.K. Chung, L. Lu, V. Vu, The spectra of random graphs with given expected degrees, *Internet Mathematics* **1** (2004) 257-275.

[80] C. Cooper, A. Frieze, On a general model of web graphs, *Random Structures Algorithms* **22** (2003) 311-335.

[81] C. Cooper, A. Frieze, J. Vera, Adversarial deletion in a scale-free random graph process, *Combinatorics, Probability, and Computing* **16** (2007) 261-270.

[82] C. Cooper, R. Klasing, M. Zito, Lower bounds and algorithms for dominating sets in web graphs, *Internet Mathematics* **2** (2005) 275-300.

[83] J.N. Cooper, Quasirandom permutations, *Journal of Combinatorial Theory, Series A* **106** (2004) 123-143.

[84] R. Diestel, *Graph theory*, Springer-Verlag, New York, 2000.

[85] R. Diestel, I. Leader, A. Scott, S. Thomassé, Orientations and partitions of the Rado graph, accepted to *Transactions of the American Mathematical Society.*

[86] S.N. Dorogovtsev, J.F.F. Mendes, A.N. Samukhin, Structure of growing networks with preferential linking, *Physical Review Letters* **85** (2000) 4633-4636.

[87] S.N. Dorogovtsev, J.F.F. Mendes, Scaling properties of scale-free evolving networks: continuous approach, *Physical Review E* **63** (2001) 1-19.

[88] S.N. Dorogovtsev, J.F.F. Mendes, *Evolution of networks: From biological nets to the Internet and WWW*, Oxford University Press, Oxford, 2003.

[89] P.A. Dreyer, Applications and variations of domination in graphs, Ph.D. Dissertation, Department of Mathematics, Rutgers University, 2000.

[90] E. Drinea, M. Enachescu, M. Mitzenmacher, Variations on random graph models for the web, technical report, Harvard Computer Science Technical Report TR-06, 2001.

[91] W. Duckworth, M. Zito, Sparse hypercube 3-spanners, *Discrete Applied Mathematics* **103** (2000) 289-295.

[92] R. Durrett, *Random Graph Dynamics*, Cambridge University Press, New York, 2006.

[93] G.F.R. Ellis, H. van Elst, General relativity and quantum cosmology, In: *Theoretical and Observational Cosmology* (Marc Lachièze-Rey ed.), Kluwer, Dordrecht, 1999, pp. 1-116.

[94] P. Erdős, Some remarks on the theory of graphs, *Bulletin of the American Mathematical Society* **53** (1947) 292-294.

[95] P. Erdős, Graph theory and probability, *Canadian Journal of Mathematics* **11** (1959) 34-38.

[96] P. Erdős, On a problem in graph theory, *The Mathematical Gazette* **47** (1963) 220-223.

[97] P. Erdős, L. Lovász, Problems and results on 3-chromatic hypergraphs and some related questions, *Colloquia Mathematica Societatis János Bolyai* **11** (1975) 609-627.

[98] P. Erdős, A. Rényi, On random graphs I, *Publicationes Mathematicae Debrecen* **6** (1959) 290-297.

[99] P. Erdös, A. Rényi, On the evolution of random graphs, *Publ. Math. Inst. Hungar. Acad. Sci.* **5** (1960) 17-61.

[100] P. Erdős, A. Rényi, Asymmetric graphs, *Acta Mathematica Academiae Scientiarum Hungaricae* **14** (1963) 295-315.

[101] L. Euler, Solutio problematis ad geometriam situs pertinentis, *Commetarii Academiae Scientiarum Imperialis Petropolitanae* **8** (1736) 128-140.

[102] M. Faloutsos, P. Faloutsos, C. Faloutsos, On power-law relationships of the Internet topology, In: *Proceedings of the Conference on Applications, Technologies, Architectures, and Protocols for Computer Communication*, 1999.

[103] A. Farahat, T. LoFaro, J.C. Miller, G. Rae, L.A. Ward, Authority rankings from HITS, PageRank, and SALSA: existence, uniqueness, and effect of initialization, *SIAM Journal on Scientific Computing* **27** (2006) 1181-1201.

[104] G.W. Flake, S. Lawrence, C.L. Giles, Efficient identification of Web communities, In: *Proceedings of the Sixth ACM-SIGKDD International Conference on Knowledge Discovery and Data Mining*, 2000.

[105] A. Flaxman, A. Frieze, J. Vera, A geometric preferential attachment model of networks, *Lecture Notes in Computer Science* **3243** (2004) 44-55.

[106] A. Flaxman, A. Frieze, J. Vera, Adversarial deletions in a scale free random graph process, In: *Proceedings of ACM-SIAM Symposium on Discrete Algorithms*, 2005.

[107] A. Flaxman, A. Frieze, J. Vera, A geometric preferential attachment model of networks, *Internet Mathematics* **3** (2007) 187-205.

[108] A. Flaxman, A. Frieze, J. Vera, A geometric preferential attachment model of networks II, In: *Proceedings of The 5th Workshop on Algorithms and Models for the Web-Graph*, 2007.

[109] A.D. Forbes, M.J. Grannell, T.S. Griggs, Steiner triple Systems and existentially closed graphs, *Electronic Journal of Combinatorics* **12** (2005) Research Paper 42, 11 pp.

[110] S. Fortunato, A. Flammini, F. Menczer, Scale-free network growth by ranking, *Physical Review Letters* **96** (2006) 218701.

[111] R. Fraïssé, *Theory of relations*, *Revised edition, with an appendix by Norbert Sauer*, North-Holland Publishing Co., Amsterdam, 2000.

[112] G. Frobenius, Uber Matrizen aus nicht negativen Elementen, S.-B. Deutsch. Akad. Wiss. Berlin, Math.-Nat. Kl. (1912) 456-477.

[113] Z. Füredi, J. Komlós, The eigenvalues of random symmetric matrices, *Combinatorica* **1** (1981) 233-241.

[114] A. Gardiner, Homogeneous graphs, *Journal of Combinatorial Theory, Series B*, **20** (1976) 94-102.

[115] E.N. Gilbert, Random graphs, *Annals of Mathematical Statistics* **30** (1959) 1141-1144.

[116] Y.V. Glebskii, D.I. Kogan, M.I. Liogonki, V.A. Talanov, Range and degree of realizability of formulas in the restricted predicate calculus, *Cybernetics* **5** (1969) 142-154.

[117] A.P. Godbole, B. Weiland, On the domination number of a random graph, *The Electronic Journal of Combinatorics* **8(1)**, 2001.

[118] Y. Gol'fand, M. Klin, On k-homogeneous graphs, *Algorithmic Studies in Combinatorics*, Nauka, Moscow, 1978, pp. 76-85. (Russian)

[119] A.S. Goldstein, E.M. Reingold, The complexity of pursuit on a graph, *Theoretical Computer Science* **143** (1995) 93-112.

[120] G. Golub, C.F. Van Loan, *Matrix Computations*, Johns Hopkins University Press, London, 1996.

[121] R.L. Graham, J.H. Spencer, A constructive solution to a tournament problem, *Canadian Mathematical Bulletin* **14** (1971) 45-48.

[122] G.R. Grimmett, D.R. Stirzaker, *Probability and Random Processes*, 3rd Edition, Oxford University Press, 2001.

[123] J.W. Grossman, P.D.F. Ion, On a portion of the well-known collaboration graph, *Congressus Numerantium* **108** (1995) 129-131.

[124] J. Guare, *Six Degrees of Separation*, Vintage Books, New York, 1994.

[125] A. Gulli, A. Signorini, The indexable Web is more than 11.5 billion pages, In: *Proceedings of the 14th International Conference on World Wide Web*, 2005.

[126] G. Hahn, G. MacGillivray, A characterization of k-cop-win graphs and digraphs, *Discrete Mathematics* **306** (2006) 2492-2497.

[127] A. Hajnal, On the chromatic number of graphs and set systems, *PIMS Lecture Notes*, University of Calgary, 1994.

[128] T.W. Haynes, S. Hedetniemi, P. Slater, *Fundamentals of Domination in Graphs*, Marcel Dekker, Inc., New York, 1998.

[129] P. Hell, J. Nešetřil, *Graphs and homomorphisms*, Oxford University Press, Oxford, 2004.

[130] C.W. Henson, A family of countable homogeneous graphs, *Pacific Journal of Mathematics* **38** (1971) 69-83.

[131] M.R. Henzinger, Algorithmic challenges in web search engines, *Internet Mathematics* **1** (2004) 115-126.

[132] Y. Hirate, S. Kato, H. Yamana, Web structure in 2005, In: *Proceedings of the 4th Workshup on Algorithms and Models for the Web-Graph*, 2006.

[133] W. Hodges, *Model theory*, Cambridge University Press, Cambridge, 1993.

[134] W. Hoeffding, Probability inequalities for sums of bounded random variables, *Journal of the American Statistical Association* **58** (1963) 13-30.

[135] W. Imrich, S. Klavžar, *Product graphs: structure and recognition*, John Wiley and Sons, New York, 2000.

[136] J. Janssen, P. Prałat, Rank-based attachment leads to power law graphs, submitted.

[137] S. Janson, T. Łuczak, A. Ruciński, *Random Graphs*, John Wiley and Sons, New York, 2000.

[138] J. Jordan, The degree sequences and spectra of scale-free random graphs, *Random Structures and Algorithms* **29** (2006) 226-242.

[139] Z. Katona, T. Móri, A new class of scale-free graphs, *Statistics and Probability Letters* **76** (2006) 1587-1593.

[140] J.M. Kleinberg, Authoritative sources in a hyperlinked environment, *Journal for the Association of Computing Machinery* **46** (1999) 604-632.

[141] J. Kleinberg, S.R. Kumar, P. Raghavan, S. Rajagopalan, A. Tomkins, The web as a graph: Measurements, models and methods, In: *Proceedings of the International Conference on Combinatorics and Computing*, 1999.

[142] J. Kleinberg, R.D. Kleinberg, Isomorphism and embedding problems for infinite limits of scale-free graphs, In: *Proceedings of the Sixteenth Annual ACM-SIAM Symposium on Discrete algorithms*, 2005.

[143] M. Krivelevich, B. Sudakov, Pseudo-random graphs, In: *More Sets, Graphs and Numbers*, Bolyai Society Mathematical Studies **15**, Springer, 2006, pp. 199-262.

[144] R. Kumar, P. Raghavan, S. Rajagopalan, A. Tomkins, Trawling the web for emerging cyber-communities, In: *Proceedings of the 8th WWW Conference*, 1999.

[145] R. Kumar, P. Raghavan, S. Rajagopalan, D. Sivakumar, A. Tomkins, E. Upfal, Stochastic models for the web graph, In: *Proceedings of the 41th IEEE Symposium on Foundations of Computer Science*, 2000.

[146] R. Kumar, P. Raghavan, S. Rajagopalan, D. Sivakumar, A. Tomkins, E. Upfal, The web as a graph, In: *Proceedings of the 19th ACM SIGACT-SIGMOD-AIGART Symposium on the Principles of Database Systems*, 2002.

[147] R. Kumar, J. Novak, P. Raghavan, A. Tomkins, On the bursty evolution of Blogspace, In: *Proceedings of the International World Wide Web Conference*, 2003.

[148] A.H. Lachlan, R.E. Woodrow, Countable ultrahomogeneous undirected graphs, *Transactions of the American Mathematical Society* **262** (1980) 51-94.

[149] A.N. Langville, C.D. Meyer, *Google's PageRank and Beyond: The Science of Search Engine Rankings*, Princeton University Press, Princeton, 2006.

[150] S. Lawrence, C.L. Giles, Searching the World Wide Web, *Science* **280** (1998) 98-100.

[151] S. Lawrence, C.L. Giles, Accessibility of information on the web, *Nature* **400** (1999) 107-109.

[152] R. Lempel, S. Moran, The Stochastic approach for link-structure analysis (SALSA) and the TKC Effect, In: *Proceedings of the 9th international World Wide Web conference on Computer networks*, 2000.

[153] T.M. Liggett, *Stochastic Interacting Systems: Contact, Voter and Exclusion Processes*, Springer, New York, 1999.

[154] N. Litvak, Y. Volkovich, W. Scheinhardt, In-degree and PageRank of web pages: why do they follow similar power laws?, In: *Proceedings of the 4th Workshop on Algorithms and Models for the Web-Graph*, 2006.

[155] L. Lovász, On chromatic number of finite set-systems, *Acta Math. Acad. Sci. Hungar.* **19** (1968) 59-67.

[156] L. Lovász, Kneser's Conjecture, Chromatic Numbers and Homotopy, Journal of Combinatorial Theory, Series A **25** (1978) 319-324.

[157] T. Łuczak, P. Prałat, Protean graphs, *Internet Mathematics* **3** (2006) 21-40.

[158] A.A. Markov, Extension of the law of large numbers to variables depending on each other, *Izvestiya Fiziko-matematicheskogo obschestva pri Kazanskom universitete* **15** (1906) 135-156. (Russian)

[159] J. Matoušek, A combinatorial proof of Kneser's conjecture, *Combinatorica* **24** (2004) 163170.

[160] C.J.H. McDiarmid, On the method of bounded differences, *Surveys in Combinatorics, London Mathematical Society Lecture Notes Series* **141**, Cambridge University Press, 1989.

[161] M. Mihail, C.H. Papadimitriou, On the eigenvalue power law, In: *Proceedings of the 6th International Workshop on Randomization and Approximation Techniques*, 2002.

[162] S. Milgram, The small world problem, *Psychology Today* **2** (1967) 60-67.

[163] M. Mitzenmacher, A brief history of generative models for power law and lognormal distributions, *Internet Mathematics* **1** (2004) 226-251.

[164] M. Mitzenmacher, The future of power law research, *Internet Mathematics* **2** (2005) 525-534.

[165] M. Mitzenmacher, E. Upfal, *Probability and Computing*, Cambridge University Press, New York, 2005.

[166] M. Molloy, B. Reed, A critical point for random graphs with a given degree sequence, *Random Structures and Algorithms* **6** (1995) 161-179.

[167] M.E.J. Newman, The structure of scientific collaboration networks, *Proceedings of the National Academy of Sciences of the United States of America* **98** (2001) 404-409.

[168] A.Y. Ng, A.X. Zheng, M.I. Jordan, Link analysis, eigenvectors, and stability, In: *Proceedings of the International Joint Conference on Artificial Intelligence*, 2001.

[169] R. Nowakowski, P. Winkler, Vertex-to-vertex pursuit in a graph, *Discrete Mathematics* **43** (1983) 235-239.

[170] A. Quilliot, Jeux et pointes fixes sur les graphes, Ph.D. Dissertation, Université de Paris VI, 1978.

[171] V. Pareto, *Cours d'Economie Politique*, Geneva: Droz, 1896.

[172] R. Pastor-Satorras, A. Vespignani, Epidemic spreading in scale-free networks, *Physical Review Letters* **86** (2001) 3200-3203.

[173] M. Penrose, *Random Geometric Graphs*, Oxford University Press, Oxford, 2003.

[174] O. Perron, Zur Theorie der Matrizen, Mathematische Annalen **64** (1907) 248-263.

[175] N. Przulj, *Graph theory analysis of protein-protein interactions*, In: Knowledge Discovery in Proteomics, (I. Jurisica and D. Wigle, eds.), CRC Press, 2005.

[176] R. Rado, Universal graphs and universal functions, *Acta Arithmetica* **9** (1964) 331-340.

[177] S. Radziszowski, Small Ramsey numbers, *Electronic Journal of Combinatorics* DS1 (2006) 60 pp.

[178] F.P. Ramsey, On a problem of formal logic, *Proceedings of the London Mathematical Society* **30** (1930) 264-286.

[179] S. Redner, How popular is your paper? An empirical study of the citation distribution, *The European Physical Journal B* **4** (1998) 131-134.

[180] W.M. Schmidt, Equations over finite fields, *Lecture Notes in Mathematics,* **536**, Springer-Verlag, Berlin, Heidelberg, New York, 1976.

[181] J.J. Seidel, A survey of two-graphs. *Colloquio Internazionale sulle Teorie Combinatorie* **17** (1973) 481-511.

[182] S.B. Seidman, Network structure and minimum degree, *Social Networks* **5** (1983) 269-287.

[183] E. Seneta, *Non-negative matrices and Markov chains*, Springer, New York, 2006.

[184] E. Shamir, J. Spencer, Sharp concentration of the chromatic number on random graphs $G_{n,p}$, *Combinatorica* **7** (1987) 121-129.

[185] J. Sheehan, Smoothly embeddable subgraphs, *Journal of the London Mathematical Society* **9** (1974) 212-218.

[186] J. Spencer, Ramsey's theorem—a new lower bound, *Journal of Combinatorial Theory, Series A* **18** (1975) 108-115.

[187] I. Stojmenovic, M. Seddigh, J. Zunic, Dominating sets and neighbor elimination-based broadcasting algorithms in wireless networks, *IEEE Transactions on Parallel and Distributed Systems* **13** (2002) 14-25.

[188] L. Stubbs, *Genome comparison techniques*, In: Genomic Technologies: Present and Future (D. Galas and S. McCormack, eds.), Caister Academic Press, 2002.

[189] T. Szele, Kombinatorische Untersuchungen über den gerichteten vollständigen Graphen, *Mat. Fiz. Lapok* **50** (1943) 223–256.

[190] A.G. Thomason, Pseudo-random graphs, *Annals of Discrete Mathematics* **33** (1987) 307-331.

[191] R. Tošić, The search number of Cartesian products of graphs, *Univerzitet u Novom Sadu. Zbornik Radova Prirodno-Matematichkog Fakulteta* **17** (1987) 239-243.

[192] T. Upstill, N. Craswell, D. Hawking, Predicting fame and fortune: PageRank or in-degree? In: *Proceedings of the Eighth Australasian Document Computing Symposium*, 2003.

[193] C. Wang, Growth deletion models for the web graph and other massive networks, In: *Proceedings of the Workshop on Combinatorial and Algorithmic Aspects of Networking*, 2005.

[194] D.J. Watts, S.H. Strogatz, Collective dynamics of 'small-world' networks, *Nature* **393** (1998) 440-442.

[195] D.B. West, *Introduction to Graph Theory, 2nd edition*, Prentice Hall, Upper Saddle River, NJ, 2001.

[196] H.S. Wilf, The friendship theorem, *Combinatorial Mathematics and Its Applications,* Academic Press, New York, 1971.

[197] R.H. Zakon, Hobbes' Internet Timeline. Accessed April 5, 2007. `http://www.zakon.org/robert/internet/timeline/`

Index

Titles in This Series

89 **Anthony Bonato,** A course on the web graph, 2008
88 **Nathanial P. Brown and Narutaka Ozawa,** C*-algebras and finite-dimensional approximations, 2008
87 **Srikanth B. Iyengar, Graham J. Leuschke, Anton Leykin, Claudia Miller, Ezra Miller, Anurag K. Singh, and Uli Walther,** Twenty-four hours of local cohomology, 2007
86 **Yulij Ilyashenko and Sergei Yakovenko,** Lectures on analytic differential equations, 2007
85 **John M. Alongi and Gail S. Nelson,** Recurrence and topology, 2007
84 **Charalambos D. Aliprantis and Rabee Tourky,** Cones and duality, 2007
83 **Wolfgang Ebeling,** Functions of several complex variables and their singularities (translated by Philip G. Spain), 2007
82 **Serge Alinhac and Patrick Gérard,** Pseudo-differential operators and the Nash–Moser theorem (translated by Stephen S. Wilson), 2007
81 **V. V. Prasolov,** Elements of homology theory, 2007
80 **Davar Khoshnevisan,** Probability, 2007
79 **William Stein,** Modular forms, a computational approach (with an appendix by Paul E. Gunnells), 2007
78 **Harry Dym,** Linear algebra in action, 2007
77 **Bennett Chow, Peng Lu, and Lei Ni,** Hamilton's Ricci flow, 2006
76 **Michael E. Taylor,** Measure theory and integration, 2006
75 **Peter D. Miller,** Applied asymptotic analysis, 2006
74 **V. V. Prasolov,** Elements of combinatorial and differential topology, 2006
73 **Louis Halle Rowen,** Graduate algebra: Commutative view, 2006
72 **R. J. Williams,** Introduction the the mathematics of finance, 2006
71 **S. P. Novikov and I. A. Taimanov,** Modern geometric structures and fields, 2006
70 **Seán Dineen,** Probability theory in finance, 2005
69 **Sebastián Montiel and Antonio Ros,** Curves and surfaces, 2005
68 **Luis Caffarelli and Sandro Salsa,** A geometric approach to free boundary problems, 2005
67 **T.Y. Lam,** Introduction to quadratic forms over fields, 2004
66 **Yuli Eidelman, Vitali Milman, and Antonis Tsolomitis,** Functional analysis, An introduction, 2004
65 **S. Ramanan,** Global calculus, 2004
64 **A. A. Kirillov,** Lectures on the orbit method, 2004
63 **Steven Dale Cutkosky,** Resolution of singularities, 2004
62 **T. W. Körner,** A companion to analysis: A second first and first second course in analysis, 2004
61 **Thomas A. Ivey and J. M. Landsberg,** Cartan for beginners: Differential geometry via moving frames and exterior differential systems, 2003
60 **Alberto Candel and Lawrence Conlon,** Foliations II, 2003
59 **Steven H. Weintraub,** Representation theory of finite groups: algebra and arithmetic, 2003
58 **Cédric Villani,** Topics in optimal transportation, 2003
57 **Robert Plato,** Concise numerical mathematics, 2003
56 **E. B. Vinberg,** A course in algebra, 2003
55 **C. Herbert Clemens,** A scrapbook of complex curve theory, second edition, 2003
54 **Alexander Barvinok,** A course in convexity, 2002

53 **Henryk Iwaniec,** Spectral methods of automorphic forms, 2002
52 **Ilka Agricola and Thomas Friedrich,** Global analysis: Differential forms in analysis, geometry and physics, 2002
51 **Y. A. Abramovich and C. D. Aliprantis,** Problems in operator theory, 2002
50 **Y. A. Abramovich and C. D. Aliprantis,** An invitation to operator theory, 2002
49 **John R. Harper,** Secondary cohomology operations, 2002
48 **Y. Eliashberg and N. Mishachev,** Introduction to the h-principle, 2002
47 **A. Yu. Kitaev, A. H. Shen, and M. N. Vyalyi,** Classical and quantum computation, 2002
46 **Joseph L. Taylor,** Several complex variables with connections to algebraic geometry and Lie groups, 2002
45 **Inder K. Rana,** An introduction to measure and integration, second edition, 2002
44 **Jim Agler and John E. McCarthy,** Pick interpolation and Hilbert function spaces, 2002
43 **N. V. Krylov,** Introduction to the theory of random processes, 2002
42 **Jin Hong and Seok-Jin Kang,** Introduction to quantum groups and crystal bases, 2002
41 **Georgi V. Smirnov,** Introduction to the theory of differential inclusions, 2002
40 **Robert E. Greene and Steven G. Krantz,** Function theory of one complex variable, third edition, 2006
39 **Larry C. Grove,** Classical groups and geometric algebra, 2002
38 **Elton P. Hsu,** Stochastic analysis on manifolds, 2002
37 **Hershel M. Farkas and Irwin Kra,** Theta constants, Riemann surfaces and the modular group, 2001
36 **Martin Schechter,** Principles of functional analysis, second edition, 2002
35 **James F. Davis and Paul Kirk,** Lecture notes in algebraic topology, 2001
34 **Sigurdur Helgason,** Differential geometry, Lie groups, and symmetric spaces, 2001
33 **Dmitri Burago, Yuri Burago, and Sergei Ivanov,** A course in metric geometry, 2001
32 **Robert G. Bartle,** A modern theory of integration, 2001
31 **Ralf Korn and Elke Korn,** Option pricing and portfolio optimization: Modern methods of financial mathematics, 2001
30 **J. C. McConnell and J. C. Robson,** Noncommutative Noetherian rings, 2001
29 **Javier Duoandikoetxea,** Fourier analysis, 2001
28 **Liviu I. Nicolaescu,** Notes on Seiberg-Witten theory, 2000
27 **Thierry Aubin,** A course in differential geometry, 2001
26 **Rolf Berndt,** An introduction to symplectic geometry, 2001
25 **Thomas Friedrich,** Dirac operators in Riemannian geometry, 2000
24 **Helmut Koch,** Number theory: Algebraic numbers and functions, 2000
23 **Alberto Candel and Lawrence Conlon,** Foliations I, 2000
22 **Günter R. Krause and Thomas H. Lenagan,** Growth of algebras and Gelfand-Kirillov dimension, 2000
21 **John B. Conway,** A course in operator theory, 2000
20 **Robert E. Gompf and András I. Stipsicz,** 4-manifolds and Kirby calculus, 1999
19 **Lawrence C. Evans,** Partial differential equations, 1998
18 **Winfried Just and Martin Weese,** Discovering modern set theory. II: Set-theoretic tools for every mathematician, 1997

For a complete list of titles in this series, visit the
AMS Bookstore at **www.ams.org/bookstore/**.